Community Vision and Leadership in Practice

This book is about building and maintaining involved, sustainable, and inclusive communities from the ground up during a period of unprecedented growth and global change. It explains the concepts and principles of community and sustainability and provides students with a framework of sustainable community planning to put into practice. It is also designed to help communities everywhere identify and reconnect the true essence of their ecological setting with the objective of raising their quality of life by increasing social, environmental, and economic sustainability.

Features:

- Provides up-to-date frameworks for sustainable community planning processes and case studies on community planning
- Explains tools for sustainable planning in accessible (non-specialist) language
- Illustrates a roadmap to an inclusive, collaborative future
- Explains aspects of sustainable community planning to maximize ecological ecosystem services and climate co-benefits simultaneously
- Includes discussion questions and suggestions following each chapter

Intended for undergraduate and graduate students taking leadership and community courses with an emphasis on sustainable practices and ethics, as well as for citizens and professionals involved in community projects related to sustainability, the authors provide a forward-thinking approach, showing readers that they are capable of making a positive impact on the future of community development through sustainable approaches and ethical leadership practices.

Social–Environmental Sustainability

Series Editor
Chris Maser

Published Titles

Decision-Making for a Sustainable Environment
A Systemic Approach
Chris Maser

Biosequestration and Ecological Diversity
Mitigating and Adapting to Climate Change and Environmental Degradation
Wayne A. White

Insects and Sustainability of Ecosystem Services
Timothy D. Schowalter

Land-Use Planning for Sustainable Development, Second Edition
Jane Silberstein and Chris Maser

Interactions of Land, Ocean and Humans
A Global Perspective
Chris Maser

Sustainability and the Rights of Nature
An Introduction
Cameron La Follette and Chris Maser

Fundamentals of Practical Environmentalism
Mark B. Weldon

Economics and Ecology
United for a Sustainable World
Charles R. Beaton and Chris Maser

Resolving Environmental Conflicts
Principles and Concepts, Third Edition
Chris Maser and Lynette de Silva

Sustainability and the Rights of Nature in Practice
Cameron La Follette and Chris Maser

Resolving Water Conflicts Workbook
Lynette de Silva and Chris Maser

Community Vision and Leadership in Practice
A Sustainable Approach
Chris Maser and Holly V. Campbell

For more information on this series, please visit: www.routledge.com/Social-Environmental-Sustainability/book-series/CRCSOCENVSUS

Community Vision and Leadership in Practice

A Sustainable Approach

Chris Maser and
Holly V. Campbell

CRC Press
Taylor & Francis Group
Boca Raton London New York

CRC Press is an imprint of the
Taylor & Francis Group, an informa business

Designed cover image: Alfama, Lisbon District, Portugal © Holly V. Campbell

First edition published 2024
by CRC Press
6000 Broken Sound Parkway NW, Suite 300, Boca Raton, FL 33487–2742

and by CRC Press
4 Park Square, Milton Park, Abingdon, Oxon, OX14 4RN

CRC Press is an imprint of Taylor & Francis Group, LLC

ISBN: 978-1-032-40287-1 (hbk)
ISBN: 978-1-032-40580-3 (pbk)
ISBN: 978-1-003-35374-4 (ebk)

DOI: 10.1201/9781003353744

Typeset in Palatino LT Std
by Apex CoVantage, LLC

Contents

Foreword

As an educator for over four decades, I have watched students light up while learning outdoor-hands-in-the-soil or seen them glaze over at the mention of dreaded paper-and-pencil assignments that remained the same over the course of their generation. My career has been spent as a champion for kindness over bullying, teaching the whole child, using technology effectively, exploring the nature we have right in front of us, and taking charge of our future by becoming community scientists. I have taught in a regular high school, a residential psychiatric treatment institution, a residential drug and alcohol treatment facility, and alternative education.

My current endeavor for the last 15 years has been to create a self-contained program for ninth- and tenth-grade students who are not gaining traction toward earning a diploma. The program is based on running a half-acre farm on our alternative education campus. We coordinate with local and state environmental organizations, Oregon State University, and agricultural experts that grow produce commercially. We raise chickens, honeybees, mason bees, salmon, and trout. We visit local sites weekly. Behind the scenes, I write two to three grants per year.

I spend my time creating community connections, bringing local experts into the classroom, and moving away from textbooks after realizing that the information base is so vast and ever-changing that a textbook would not allow us the freedom to investigate the questions that need answers right now. In addition, textbooks do not help us know our particular surroundings. That which is native to our area is personal and sacred.

Chris Maser (the lead author of this book) and I have been thrown together in the most grassroots way imaginable. Even though I live 20 miles away from him, I consider Chris my closest neighbor. Chris and his wife, Zane, live across the street from the school where I have put into practice my reworking of education throughout the course of my many years of teaching. Chris has watched me gradually move more of my teaching outdoors and finally to the creation of the Urban Farm Program. He and I have had numerous "over the fence" conversations about youth and educational trends.

Chris is insightful, eloquent, and humble and is a great listener, and for reasons I still do not understand, he is patient. How can anyone who has clearly discovered the key to a sustainable society and the formula for true leadership not be screaming at the top of his lungs, "I have the answer. It's simple and it's right in front of us"? It is through the many publications he has written, the trainings and interventions, and finally an entire website dedicated to his life's work that helps us to feel heard and validated.

There are some books we read that make us want to highlight passages and dog-eared pages and then beg specific people to read the book. *Community Vision and Leadership in Practice: A Sustainable Approach* is that type of book. This book, written by Chris and Holly, is perfect for groups to read, gather, and discuss. I zeroed in on leadership and education, but any governing body or large business would benefit from this title. It is past time to reorganize our priorities and provide a planet that guarantees the children of today their right to clean air and drinkable water. How can any decisions be made that do not take this priority into consideration? If it hurts the planet or hurts any living being, it must be *carefully* considered before it is implemented, and all vested parties must agree to it.

When Chris asked me to write this foreword, I told him that I am nobody special and that someone well known should write it. Chris said, "I have a gift and so do you. I want you to speak through yours and be heard." And this, my friends, is how a true leader leads. This book is a pat on the back and a shove in the right direction. It can easily be applied to any level of organization: parenting, teaching, running a small business, or leading a large corporation. It is written in an easily consumed format, and I believe that everyone would benefit from reading this book. Our systems that loosely keep humanity functioning need to be reworked from the ground up. Either we give in, or the environment gives up. This book is for everyone.

Cherie Taylor
Creator and Lead Teacher of the Urban Farm Program
College Hill High School
Corvallis, Oregon

Preface

> The most intense conflicts, if overcome, leave behind a sense of security and calm that is not easily disturbed. It is just these intense conflicts and their conflagration which are needed to produce valuable and lasting results.
>
> —**Carl Gustav Jung**[1]

As we are drawn ever-closer together through technological advances, we may sense our communities are more fragmented on one hand yet more interdependent on the other. We are now a global community of eight billion and counting! The evolving awareness of our interconnection and diversity poses fundamental challenges to old ways of thinking, communicating, and behaving within and among cultures. How we respond to the mounting challenges as individuals and communities will determine whether we become truly nurturing, cohesive, progressive, and sustainable or inhospitable, divided, and resistant to change—the latter risking continually feeding the imploding global crisis of diminishing sustainability.

We, the authors, have worked with vision statements, goals, and objectives for many years. Our approach to helping people formulate collective visions has been through the medium of resolving environmental conflicts, where a statement of vision and its attendant goals are committed to paper, as the culmination of the resolution process. In this sense, the frequent chaos of resolving a conflict is actually a prelude to creativity in which people are willing to craft and implement a shared vision toward which to build as an alternative to the continuing uncertainty of conflict. In some situations, people may ultimately be willing to implement a group vision as a means of avoiding further pain—but only as a last resort.

We have occasionally taken part in so-called community visioning processes in which it was patently clear that the people—at all levels of government—conducting the process knew nothing about a vision or how to create one. Under such circumstances, while collective "visions" have been forthcoming, animated by the people's desire—often an agenda of city government—for a guaranteed, self-determined outcome, they have not been implemented. The question is why. Was it to protect a sense of control and hence maintain the status quo? Was it because people did not really care about the future (or that of their children and grandchildren) so long as they knew how to obtain what they wanted in the present, and perhaps into the perceived, foreseeable future?

If there was no sincere, compelling reason to craft a shared vision in the first place, why did communities go to the trouble and expense to do so, only to abandon the process in the implementation stage? Was it a lack of true leadership, the required expense of time or funding, or was it the new set of uncomfortable social constraints for which the citizenry would hold city employees accountable if the vision were to be achieved?

In the years we have dealt with government and communities, from the local to the state and federal levels, our experience has been largely one of concerted resistance to change, even in the face of a progressive and undeniable decline in our quality of life, which raises a curious condition of human idealism, aptly expressed by John Galsworthy: "Idealism increases in direct proportion to one's distance from the problem."[2]

A community, state, or nation is only as good as the home life of its people. In essence, our national government is too big for the small, local, community-based problems of life and too small in bipartisan spirit for the big problems. Local communities must therefore redouble dedication to their own care and their own future, which means they must always make full use of a good mistake, which brings us to the notion of leadership. Ideally, our governance structures must be nested or networked, well coordinated, and collaborative.

It has been our experience, both at home and abroad, that the vast majority of people cannot lead because they do not understand the inner qualities of leadership, let alone how to create and implement a viable shared vision for the sustainable future of a community, which demands the utmost skill. For this reason, the present book is not intended as a how-to book because there already exists a plethora of intellectual/mechanical models, especially for creating a shared vision and goals. Instead, this book is intended to probe the seldom-articulated philosophical underpinnings of these models and examine the qualities of leadership—the language of the heart—that it would take to implement them.

Throughout this book, we define the concept of sustainability in keeping with the definition in the Brundtland Commission Report. Thus, in this book sustainability means "meeting the needs of the present without compromising the ability of future generations to meet their needs."[3] We also refer throughout to the three pillars of sustainability: ecological, social, and economic.[4]

References

1. Carl Jung. www.brainyquote.com/authors/carl-jung-quotes (accessed January 13, 2022).
2. John Galsworthy. www.brainyquote.com/quotes/john_galsworthy_161985 (accessed January 13, 2022).
3. Our Common Future. 1987. https://sustainabledevelopment.un.org/content/documents/5987our-common-future.pdf (accessed March 4, 2022).
4. Ben Purvis, Yong Mao, and Darren Robinson. 2019. Three Pillars of Sustainability: In Search of Conceptual Origins. *Sustainability Science*, 14:681–695.

1

Why Is a Shared Vision Important?

1.1 Introduction

> The good we can do together surpasses the good we can do alone.
>
> **—Benjamin Franklin**[1]

Stories of the universe and the Earth have been told by many peoples since the beginning of humanity. These stories are celebrated in elaborate rituals that have shaped us and provide guidance and sustaining energy. The stories are at the root of our societies establishing social authority regarding the modes and bounds of personal and community conduct.

We, in the modern period, are without a comprehensive story of the Earth and the universe. Historians, even when articulating world history, deal not with the whole of the world but just with the human aspect, as though the human aspect is something apart from or an addendum to the story of the Earth and the universe.

Detailed scientific accounts of the cosmos focus exclusively on the physical dimension and ignore the human dimension of the story. We in the industrialized nations, with all our learning, knowledge, and scientific insight or vision, may sense we are psychologically and spiritually adrift when it comes to a meaningful relationship with our environment and the universe reflected in worldwide Indigenous creation stories. Thus, cast adrift, we experience a distorted (Buddhists might use the word delusional) approach to living on our home planet.[2]

1.2 Working Together for a Common Future

In a saying commonly attributed to Jonathan Swift, eyesight is nothing without vision, and vision is the art of seeing the invisible.[3] To this, Dag Hammarskjöld adds, "Only he who keeps his eye fixed on the far horizon [the vision] will find his right road."[4] A collective vision is founded on the

DOI: 10.1201/9781003353744-1

fact that we have the vision in common, that it is shared experience. If we fail to understand this foundation, we are unconsciously risking our collective future. When asked about their future, a person may say, "I have an idea, a vision, of what I want." But when pressed to explain their aspirations, it becomes clear they have not the slightest understanding that a vision is a strong, organizing framework built on cooperative, systemic, social-environmental interrelationships.

The organizing framework of a shared vision has distinctive traits: (1) it tends to focus on a wide range of human concerns; (2) it is strongly centered in the community; (3) it can systematically investigate and even role-play alternative scenarios to explore various realistic pathways toward the goals of the community; (4) the framework's creation relies on the trust, respect, and inclusivity of interpersonal relationships; (5) it is ideally suited to and depends on public involvement; and (6) the use of creative, graphic imagery is ideally suited to conveying the options the community develops and deliberates. In regard to land-use planning, although a shared vision does not replace existing planning tools, it is the organizational context within which all other planning activities fit, a context that a community might have developed long ago but is too often forgotten or even ignored.

More than 3,000 years ago, sages gave the whole world one of the earliest spiritual treasures known to history, the Rig Veda. It is a prayer addressed to all humanity, a prayer that is the heart and soul of a shared vision:

> Meet together, talk together:
> May your minds comprehend alike:
> Common be your action and achievement:
> Common be your thoughts and intentions:
> Common be the wishes of your hearts
> So, there may be union amongst you.
>
> Let your aims be common, and yours hearts of
> One accord, and all of you be of one mind, so
> you may live well together.[5]

A vision consists of the self-determination that people ideally want to move toward—not what they fear and want to move away from. As such, a vision, like the social-environmental sustainability of a community, is a perpetual work in progress. Always the price of a vision is courage because, at its center, it is about consciously doing the best we can to honorably push the limits of human possibility, a notion that is severely tested in times of turmoil. Pushing these limits is a necessary condition of social evolution. As the aphorism popularly attributed to Albert Einstein goes, "No problem can be solved from the same consciousness that created it."[6]

A shared vision of a sustainable future toward which a community can build creates confidence, consensus, and energy in equal parts.[7] At a deeper level, it engages our imagination and helps to ferret out which questions need to be asked, how to word them, and when to ask them. It has the power to draw us together, closer.

By engaging their imagination and sense of possibility of the ideal through countless small-scale initiatives, such as shared community visions, people who are concerned with the health of their environment and social justice can create an opportunity to confirm a more positive, sustainable future. "Imagination," as Albert Einstein said, "is more important than knowledge,"[8] but imagination without the tenacity of purpose and persistence of action is all but useless, as noted by President Calvin Coolidge:

> Nothing in the world can take the place of persistence. Talent will not; nothing is more common than unsuccessful [people] with talent. Genius will not; unrewarded genius is almost a proverb. Education will not; the world is full of educated derelicts. Persistence and determination alone are omnipotent. The slogan "Press On," has solved and always will solve the problems of the human race.[9]

Some people will point out that persistence and determination may have worked in the past because the pace of social change was slower and easier to assimilate into cultural norms. They might also contend that we barely have time to catch our breath before another whirlwind of change descends upon us. This notion may lead them to abandon notions of responsibility to each other and to the future and excuse their respective communities for choosing to operate in a perpetual, short-term thinking or crisis mode, seldom getting beyond the next budget cycle—let alone creating a shared vision for a sustainable future through which change can be wisely accommodated.

As local economies are being regionalized, nationalized, and even globalized through the expansion of corporate economic power, some industries that were once the bedrock of communities are marginalized, as new industries and technologies offer a different vision of the future. In addition, many communities are being affected by shifting demographic trends (such as changes in the work force and increasing cultural diversity) and shifting social essentials of their citizens, requiring an intelligent and inclusive reexamination of social services and their infrastructure.

Sometimes people (and politicians) express the need to be "proactive" or "get ahead of change." Yet too often they rarely recognize change for what it often is: an ongoing process of everything irreversibly becoming something else; a process that is possible to embrace for the opportunities (not just the uncertainties) present—a process that can be both thoughtfully led and deftly guided for the community's sustainable benefit.

When we cling to aversion to change, we may continually avoid uncomfortable situations until it is too late to effectively engage them. If, however, we want to live in a future of improved social equality, we must examine our own materialistic values and habits—and be determined and persistent in change as growth. Each of us in the human community must invest thought and time into becoming physically *and* psychologically prepared for the opportunities and responsibilities that are emerging as a result of our growing interdependence.

People need to develop the attitudes, values, knowledge, and skills necessary to participate confidently and constructively in shaping their own communities on all levels so that each community might reflect into the global society the principles of love, justice, equality, unity, and social-environmental sustainability for all generations into the unforeseeable future. Here, education is indispensable and must encourage thinking in terms of historical processes, seeing in history the inexorable movement toward a world civilization of ever-increasing consciousness, a movement whose successes—and failures—are the legacy of all generations.

A shared vision of a sustainable future must therefore be both ecologically and economically sound if it is going to enable people to think differently about their lives and in so doing commence to change them, but that change must be for and about the people themselves. If it does not have a human face out of which shine core human values, it will fail.

Indeed, history is replete with failed plans that omitted the human equation. The longer disorganization or dysfunction exists, the more difficult it is to reverse the effects. Yet as American author Van Wyck Brooks observed, "Nothing is so soothing to our self-esteem as to find our bad traits in our forebears. It seems to absolve us."[10]

A vision can also be likened to an old parable from India about an elephant and the blind people who describe the elephant as they are feeling it. One describes a thick pillar, another a rope, and the third a hard, curved horn. Although this story may be familiar, the Sufi version has an interesting twist. The people are not blind but rather are in a dark room. Thus, when they all light their candles, they can see the full shape, dimension, and movement of the elephant. This story depicts the point that no one person can grasp the breadth and depth of the changes that are permeating our culture today—it requires the light of many candles burning simultaneously to perceive the whole.[11]

My (Chris') experience has been that those people in a community who tend toward psychological maturity light the candles by speaking for the children (present and future), whereas those who tend toward psychological immaturity speak in the dark only for themselves. With the former, sustainability is possible. With the latter, it is not—because the children have no voice.

Knowing intuitively that children need a voice, it was an ancient custom of the Indigenous Americans to call a council fire when decisions that would affect the whole tribe or nation needed to be made. To sit on the council as a representative of the people was an honor that had to be earned through many years of truthfulness, bravery, compassion, sharing, listening, justice, being a discreet counselor, and so on. These qualities were necessary because a council fire, by its very nature, was a time to examine every point of view and explore every possibility of a situation that would in some way affect the whole of the people's destiny. In today's terminology, this might be called a visioning process.

Borrowing from these practices, the modern visioning process is designed around the same timeless teaching of the ancient council fire—namely, that until *all* of the people are doing well, *none* of the people is doing well. Thus, for the sake of sustainability, present and future, the good of the whole must be placed before the good of the few.

The council fire worked well for the First Americans because they knew who they were culturally, and they had a sense of place within their environment. Today, however, societies around the globe are in transition, which robs many people of their original sense of place and substitutes some vague idea of location. With so many people feeling adrift in their lives, the whole concept of a vision, much less one that includes future generations, may seem inconsequential at best if not an exercise in abject futility.

This transition is largely the result of massive shifts in human populations over the last three centuries. These shifts have altered the composition of peoples and their cultural structures throughout the world. All of this activity emphasizes our need for a growing awareness and acknowledgment of our interconnectedness and interdependence in order to confront cultural uncertainty.

Such uncertainty may be more acute for those who are caught between two cultures, such as the warring religious factions around the world, where millions of refugees not only have their sense of culture disrupted but also their sense of place transformed into an alien location. A large number of these people are immigrating to industrialized nations, where many are trying to reconstruct the foundations of their own cultures while fitting compatibly into a new culture and a new community.

These shifts in population are forcing even some of the most parochial communities to see themselves from new and different points of view. Others are being forced to look at themselves anew because today's social-environmental-economic conditions are changing so fast that many of our known and comfortable assumptions are obsolete. Hence, there are some preliminary subjects, beginning with our own perception of our community and world, that need examination prior to dealing with the notion of a vision itself.

1.3 References

1. This quote is attributed to Benjamin Franklin in a documentary about him on the Oregon Public Broadcasting System (accessed April 18, 2022).
2. The Preceding Three Paragraphs Are Based on: Brian Swimme and Thomas Berry. 1992. *The Universe Story: From the Primordial Flaring Forth to the Ecozoic Era—A Celebration of the Unfolding of the Cosmos.* HarperCollins, San Francisco, CA., 305 pp.

3. Jonathan Swift, as quoted at www.goodnewsnetwork.org/jonathan-swift-quote-on-vision/ (accessed January 16, 2022).
4. Dag Hammarskjöld, Markings, Knopf, 1964, ISBN-10 039443532X, ISBN 13 978–0394435329.
5. Abhilash Rajendran. *Teachings from Rig Veda—A Collection of Teachings and Wisdom of Rig Veda.* www.hindu-blog.com/2019/05/teachings-from-rig-veda-collection-of.html (accessed January 16, 2022).
6. Albert Einstein. As Quoted in All Albert Einstein Quotes about "Consciousness." www.inspiringquotes.us/author/3804-albert-einstein/about-consciousness (accessed January 16, 2022).
7. Ed Mayo and Edward Hill. 1966. Shared Vision. *Resurgence*, 177:12–14.
8. Albert Einstein. 1929. What Life Means to Einstein: An Interview by George Sylvester Vierick. *The Saturday Evening Post* (October 26, 1929). www.saturdayeveningpost.com/wp-content/uploads/satevepost/what_life_means_to_einstein.pdf.
9. Calvin Coolidge. 1932. The Calvin Coolidge Presidential Library indicates the quote occurred when Coolidge was a board member of New York Life Insurance Company, then turned into a poster for the employees. "People" is substituted for "men" by the authors. See: https://forbeslibrary.org/blog/2008/12/10/favorite-calvin-coolidge-quotes/.
10. Van Wyck Brooks. 1958. Family Systems Theory. *Essay.*
11. As quoted in Sarah van Gelder. 1997. Beyond Greed & Scarcity. *YES! A Journal of Positive Futures*, Spring:34–39.

1.4 Discussion Questions

1. What is meant by a shared vision?
2. Why is a shared vision critical with respect to community sustainability?
3. Why is it necessary to accommodate nature's processes when crafting a community vision for long-term sustainability?
4. How would you explain a shared vision for sustainability to a group of community leaders and interested members who had no idea of what social-environmental sustainability meant?
5. How would you simultaneously balance the people's social desire for continual economic growth and prosperity and the biophysical limitations of nature's processes to ensure environmental sustainability for all generations—present and future?

2

The Concept of Community

2.1 Introduction

"Civilization," as English historian Arnold Toynbee said, "is a movement and not a condition, a voyage and not a harbor."[1] The word community comes from the Latin communitas: Public spirit. Thus, community can be thought of as a group of people who welcome, honor, and share one another's gifts, burdens, duties, and obligations.

This central core of community is then expanded to a group of people with mutual interests living under and exerting some influence over the same government in a chosen locality. To understand the limitations of this notion of community, consider an extended family. The strongest bond is between a married couple or between domestic partners, then between parents and their children. But as the family grows, the bonds among the children and the various aunts and uncles and first, second, and third cousins may become progressively weaker, as relationships become more distant in both time and space.

Thus, with each higher level of organizational complexity, there is an increase in the size of the system and a corresponding decrease in the energies, or closeness of personal relationships holding it together. Consequently, with the growth of a family or town, the bonds of personal relationships—of community, if you will—can weaken as the size of the system increases; or a community in a large city may be made up of cohesive neighborhoods. This is why the concrete notion of community was largely confined to a local environment and was not conceived to extend beyond the local without becoming an untenable abstraction. However, are we not increasingly aware of the systems view—that we are part of the global community, the planetary community? What does membership in our larger human collective entail? What can we do to extend respect and responsibility to our community worldwide? Our community itself can grow (and must grow) from our local roots to the leaves and branches of the larger tree if we are to thrive amid change.

President Lincoln expressed a grasp of the government's role with respect to the community: "The legitimate objective of government is to do for the people what needs to be done, but which they can not, by individual effort, do at all, or do so well, for themselves."[2]

DOI: 10.1201/9781003353744-2

Lincoln's "legitimate objective of government" addresses local problems first and foremost with local resources. Its intent is to keep local earnings circulating within a community for as long as possible. In addition, it requires developing and practicing new forms of full-cost accounting that will result in least-cost planning so that local government is, to the maximum extent, the highest level of government needed to solve a community's problems.

People in a local community also share social interactions with one another and organizations beyond government and through such participation are able to satisfy the full range of their daily requirements within the local area. However, communities are unlikely to be prosperous or sustainable until the spiritual dimension of human reality is accounted for in the culture. This notion brings to mind a statement by German-born missionary and Nobel laureate Albert Schweitzer: "Man must cease attributing his problems to his environment, and learn again to exercise his . . . personal responsibility in the realm of faith and morals."[3]

A community also interacts with the larger society, both in creating change and in responding to it, if the change is acceptable, or reacting to it, if it is not. Finally, the community, as a whole, interacts with the local environment, molding the landscape within which it rests and in turn is molded by it. In this sense, community is about the oneness of the whole and the wholeness of the one.

Although humankind has been concerned about the health or wholeness of communities since the earliest times, that emphasis has shifted in recent years to a broader model of community health and long-term sustainability. Today's view reflects a modern understanding of the determinants of both environmental and social health, including the spiritual, behavioral, political, economic, biological, and medical aspects of society.

We are finally recognizing that health is more than the absence of disease. A comprehensive framework of health that, beyond prevention and treatment of disease, emphasizes a life filled with meaning, connection, and joy; equity; lifestyles conducive to both mental and physical health; and a comprehensive, integrated, sustainable environment.

For perhaps the first time in human history, professional experts from different disciplines are coming to an agreement with one another. The best and the brightest people from the physical and mental health professions, science, agriculture, the conservation movement, and both Eastern and Western religions are urging people to recognize that we are part of—partners with and within—a living system.

Every thought, every word, and every action creates a response that affects the health (ease) or disease of the whole system. The power abiding in our individual conscience and our collective consciousness presents our greatest potential to affect change. It is from an integrated, "whole system" perspective that together we can realize the possibility of sustainable community development.

2.2 True Community Is Founded on a Sense of Place, History, Respect, and Trust

Community is rooted in a sense of place through which people are in a reciprocal relationship with their landscape. As such, a community is not simply a static place within a static landscape but rather a lively, ever-changing, interactive, and interdependent system of relationships. Because a community is a self-organizing system, it does not simply incorporate information but acts as a self-reinforcing social-environmental feedback loop, meaning that, as a community in its living alters the landscape, the landscape in its response alters the community.

Such reciprocity either extends sustainability to or withholds it from a community and its landscape. We, therefore, create trouble for ourselves in a community when we confuse order with control. Although freedom and order are partners in generating a viable, well-ordered, autonomous community, a community is nevertheless an open system that resists continual change that is disruptive.

A person who is mindful and feels a deep sense of place can bring about change that is beneficial to the continual health of his or her community. But if, after some time, a person is told by the company he or she works for that a transfer is likely within two years, that person may go from being a resident with a sense of place to a transient reluctant to invest any more money or time in his or her home, which must be sold. As more people are put in such a position, the kinds of changes that take place in a community can strain the community's sense of its own history.

History reflects how we see ourselves and how we give value to things across and within cultures. Our vision of the past is shaped by and, in turn, shapes our understanding of the present—those complex and comprehensive images we carry in our heads by which we decide what is inherently true or false.[4]

A community's history must therefore be passed from one generation to the next if the community is to know itself throughout the passage of time. In this sense, it is no small irony that during what might be the greatest era of prosperity in the United States—the decades following World War II—only the cheapest buildings were constructed. You can see the difference by comparing any richly designed post office or city hall built at the turn of the 19th century with its modern, dreary, concrete-box counterpart.

When the United States was a far less wealthy nation, things were built to endure because it would have seemed foolhardy or immoral, in our great-grandparents' day, to throw away hard-earned money and honest labor on something likely to disintegrate within 30 years.

Many buildings erected in those earlier days paid homage to history in their design, including elegant solutions to the age-old problems posed by the cycles of weather and light. They paid respect to the future because they were consciously built to endure the historical corridor of time beyond the lifetimes of the people who designed and constructed them. This level of consciousness put us in touch with the ages and connected us with a sense of continuity, a sense that we are part of a society and a history that is much larger than ourselves.

The notion of a historical corridor of time suggests that the community we help to compose cares about us and that we, in turn, must respect ourselves and all generations that will follow us, just as those who preceded us respected those who followed them. This notion is at once humbling and exhilarating. Connectivity with the civilizations of the past and the horizons of the future lead us in the direction of enchantment, grace, and balance.

If, however, the historical continuity of a community is disrupted, the community may suffer an extinction of identity and no longer view its landscape as an inseparable extension of itself but rather as a separate commodity to be exploited for profit. When this happens, community trust may become strained in the face of growing competition from outside enterprises with exploitive means and intentions.

Community can also be lost when there is a failure to balance endless development—construction and expansion—with the quality maintenance of created infrastructure. In today's world, some people seem eager to build but do not simultaneously balance the dollars for construction with those necessary for maintenance. Some townspeople (citing stagnant wages) may be reluctant to spend money on maintenance despite the continual increases in annual property taxes for the very purpose of maintenance of the ever-growing and aging town. In addition, in the United States, particularly during the 1980s and 1990s, some historic downtowns were essentially abandoned due to peripheral expansion. With new generations' contemporary interest and skill in urban reimagining, revisioning, and renewal, this phenomenon has finally turned around in many places.

What we neglect we lose, be it a house, a street, or a downtown. Communities are not made to be disposable; they are not designed in terms of planned obsolescence. If each of us as a member of a community would contribute something—be it time, participation, volunteering, or a beautiful garden, that contribution would improve the quality of the community. Donating (recast it as investing) a portion of one's time to the community is the beginning of recognizing intrinsic value, the difference between wealth and money.

It seems clear, therefore, that the heart of true community begins in a local place, mutual caring, and history. Community, says Wendell Berry, "is an idea that can extend itself beyond the local, but it only does so metaphorically. The idea of a national or global community is meaningless apart from the realization of local communities."[5]

We can recognize and respect that truth for Wendell Berry, who was self-admittedly (happily) place-bound on his tiny family farm in 1993 while simultaneously recognizing that his is a place of privilege not everyone is free to enjoy and that all local communities are, in fact, interdependent and connected to the larger world. We can—we have the innate capacity and imagination to—begin feeling a kinship with communities everywhere, not just at home. Maybe community, as a bedrock human concept, is not just an abstraction the farther out in scale we look. We can, if we dare, feel we are one. In any event, for a community to emerge in the first place, prevail, and be sustainable, it must be composed of trust, and trust takes much time, work, and patience.

> [A] community does not come together by covenant, by a conscientious granting of trust. It exists by proximity, by neighborhood; it knows face to face, and it trusts as it knows. It learns, in the course of time and experience, what and who can be trusted. It knows that some of its members are untrustworthy, and can be tolerant, because to know in this matter is to be safe. A community member can be trusted to be untrustworthy and so can be included. But if a community withholds trust, it withholds membership. If it cannot trust, it cannot exist.[6]

It is important here to understand that communal trust is built *gradually* through levels of communication, which means people must invest the time to get to know one another.

In sum, community is relationship, and meaningful relationship is the foundation of a healthy, sustainable community. In this connection, Ralph Waldo Emerson penned the following: "It is one of the most beautiful compensations of this life that no man can sincerely try to help another without helping himself."[7]

Community also reminds one that the scale of effective organization and action has always been small local groups. As anthropologist Margaret Mead said, "Never doubt that a small group of thoughtful, committed citizens can change the world; indeed, it is the only thing that ever has."[8]

In Western industrialized societies, our misplaced expansionist, economic worldview whereby property and the boundless appetite for economic growth take the place of meaning or spirituality, may explain our loss of community. If we are to reconnect a sustainable human society with the environment, we must deeply and thoughtfully reconsider our priorities, including our sense of local community, in order to balance the material with the spiritual, our place as individuals within the whole collective system.

The aesthetics of relationship in scales of time is illustrated in a story about the replacement of the gigantic oak beams in the ceiling of one of the dining halls at Oxford University in England. University officials were concerned that they would not be able to find lumber large enough and strong enough to replace the worn-out beams. But the replacement beams were of the same quality as when the hall had been built 500 years earlier. How could this be?[9]

"Simple," explained the university's forester. "When the dining hall was originally constructed 500 years ago, our predecessors were thoughtful enough, considerate enough, and farsighted enough to plant a grove of oak trees so that the university could, when necessary, replace the beams with others of the same quality."[10] By planting the oaks all those centuries ago, one could say that the people who planted them not only understood and appreciated time in its various scales but also may well have seen time as a mystery to be contemplated, rather than a foe to be vanquished. Thinking across scales, including time, is systems thinking.

2.3 Local Communities under Stress

Throughout a history that made us more resilient (including two world wars, the Great Depression, social and political upheavals, and two pandemics—1918 and 2020)—perhaps we have not invested enough to enhance the quality of relationship—the trust that holds traditional American communities together. In the United States, even more must be done to nurture the concept and the reality of multiracial communities. Trust is the social capital that enables people to work together and commit to common causes. And there is yet another value that we have all but lost—the spiritual and practical relationship with our environment that committed humanity to honor and protect the natural systems on which we depend.

The impact of the loss of this spiritual connection is compounded by our ignorance of its absence—we do not know what we are missing. Having said this, it is clear to me (Chris) that relationships of high quality and integrity—to each other and to our ecological setting—are absolutely critical to the success of a community in translating its cultural identity into a shared, foundational vision of the future.

For a community to fulfill its vision, it must be grounded in personal ethics, which translate into social ethics. This puts the responsibility for each of our conscience and behavior where it rightfully belongs—squarely within us.

For some, community may simply be a useful, new concept to wrap around old ideas and institutions, while for others it might seem a new frame of reference about how and why people relate to one another and to the wider world and work together to enhance both. The value of community lies in building a bridge between people's core values and the principles for action and governance. The bridge will help shift perceptions of the true purposes of politics and government: nurturing balanced growth and fulfillment of a community with its biophysical sustainability.

One relevant question regards how a community can shape and even limit its physical growth—its footprint. The decision is then written into the city charter and used to provide a context for further planning and building.[11]

The tool commonly used is called an urban growth boundary. Zoning, in conjunction with the overlay of a comprehensive land-use plan, is also critically important to designing a functional, beautiful community that accommodates thoughtful, planned growth instead of haphazard growth of hasty, ad hoc, poor development choices. The community revisits and revises these touchstone documents from time to time to ensure the documents continue to meet their needs while protecting the biophysical sustainability of the surrounding landscape for future generations.

The ability to periodically amend the city charter is important to respond to a need for more updated, defined, and authentic ethical values (trust) with which to rekindle engagement, meaning, and purpose in life and politics. A city of any size can have a form of a mission statement all its own, shaped by the citizens and leaders. Increasingly, cities are crafting what can be called "we believe" statements that reflect detailed, modern attitudes that go well beyond a traditional mission statement (what we're doing and where we want to go) to elaborate on priorities: *what we care about and why*.[12]

Community may refer to neighborhoods or workplaces. To be meaningful, however, it must imply membership in a human-scale collective, where people encounter one another face-to-face. Community must therefore nurture human-scale structural systems within which people can feel safe and at home in a particular place to which they feel a measure of fidelity. It is precisely this sense of safety in and fidelity to a place that is being called into question as the face of community is being redefined in a more-worldly context.

Using the concept and language of community is a way to reconnect people to a place and a set of shared values and principles with which to embrace daily uncertainties. Shared ethics must therefore be nurtured as one of the most valuable assets in making human communities work.

2.4 Community Depends on How We Treat One Another

2.4.1 First Americans

Prior to the arrival of Northern Europeans to the United States, the First Americans had abundant natural resources per capita on a long-term basis, although such resources as food may have been seasonally limited.

When the Europeans arrived and began competing for those same resources, the inevitable outcome was not apparent. But as the numbers of Europeans increased, through local births, rapid immigration, and the slaying of First Americans, the Indigenous peoples were increasingly pushed out of the way by superior numbers of foreigners, who demanded—and depleted—available resources.

While the various Indian tribes were dispersed enough to "share" the land, the Europeans took increasingly forcible and exclusive ownership of the contiguous whole. Moreover, they set about destroying every Indigenous culture and the sacred traditions upon which they were based by consciously superimposing their own customs and accompanying beliefs and behaviors. With the demise of their resources and traditional way of living, the First Americans faced a devastating loss of their sense of place, their many languages and sacred rituals, and the ultimate risk to their enduring sense of community and cultural identities.

2.4.2 "Company" Towns

Today, local communities may sense they are in jeopardy, as they experience increasing pressure from enormous social shifts with the continual shifts in global economics and changing technology. The capitalist system's myopic focus on the profit margin does not prioritize people, communities, cultures, or local places.

When the corporations in local communities decide to withdraw their presence, because the resources on which they count become depleted or markets fail, communities that were built around the corporations are left to fend for themselves. This often means they must use all available resources in their local landscapes because they failed to previously diversify and retool enough to survive, much less to thrive under change and upheaval.

2.4.3 Competition

In contrast to the preceding, there are communities where more privileged people may move in, drive up land prices and taxes, and effectively take over the town by economically forcing out the community's original inhabitants. The displaced members must then live elsewhere but are allowed to commute from their new homes to their original community, where they may work to serve the newcomers. Whether this happens by default or by design, the effect is the same: trust is irrevocably broken, as is the historical continuity, personality, and cohesiveness of the community. Such abrupt changes occurred in numerous places in the western United States, frequently coinciding with periods of affluence, such as the 1990s, and affecting those locations for generations. The people raised in areas affected by this kind of growth can seldom afford to settle where they were born and/or grew up.

Many people would say that competition is "simply business and not personal," but a vision directed solely by competition cannot long endure; it must deplete itself. The continual depletion of natural resources and, with them, local communities is a danger we daily face because we are so obsessed with competition that it is arguably our predominant model for learning, social communication, and change.

While there is nothing intrinsically wrong with competition—it can even be fun and promote invention and daring—we have lost the balance between competition, cooperation, and coordination at precisely the time we most need to work cooperatively with one another. Economic competition, which today is globalized, increasingly pits workers in each enterprise against workers in all enterprises, workers in each ethnic group against workers in all ethnic groups, and workers in virtually every country against workers in all countries.

Economic competition, as practiced in the utilitarian 20th-century mode, is destructive rather than constructive to social-environmental sustainability. Thus, we find ourselves oftentimes competing with the very people with whom we need to collaborate, which frequently leads to debilitating conflicts over the way in which resources are used and who gets what, how much, and for how long.

Our challenge, therefore, is to redesign and develop our communities around such universal principles as love, honesty, moderation, humility, hospitality, justice, and inclusive unity—all of which promote social cohesion. Without these ingredients, no community, no matter how economically prosperous, intellectually endowed, or technologically advanced, can long endure.

Communities will thrive and be sustainable only to the extent that they acknowledge, respect, and nurture the spiritual dimension of human nature by making the moral, emotional, spiritual, and intellectual development of the individual a central priority. Further, their centers of culture, history, and learning must cultivate equal opportunity—the limitless potentialities of the human consciousness—and must pursue the equal participation of *all* people in generating and applying their unique gifts of knowledge. This equal opportunity would include a firm, concrete commitment to safe, affordable housing, food security, accessible and affordable health care, and other necessary amenities.

Remembering at all times that the interests of the individual and the community are inseparable, not only from one another but also from the landscape within which they reside, these communities must promote respect for both the rights and responsibilities that foster equality and partnership between men and women to protect and nurture families. They must also nurture both natural and cultural beauty and use such beauty to sustainably stitch the community into its surrounding landscape.

2.5 References

1. Arnold J. Toynbee, as quoted in *Reader's Digest* (October 1958).
2. Abraham Lincoln. 1854. Fragment: On Government. *The Complete Works of Abraham Lincoln*, 2:182–183. Northern Illinois University Digital Library. https://digital.lib.niu.edu/islandora/object/niu-lincoln%3A34939.

3. Albert Schweitzer, as quoted in Goodreads.
4. www.goodreads.com/quotes/150121-man-must-cease-attributing-his-problems-to-his-environment-and (accessed January 18, 2022).
5. Georg Feuerstein, Subhash Kak, and David Frawley. 1995. The Vedas and Perennial Wisdom. *The Quest*, 8(4):32–39, 80–81.
6. Wendell Berry. 1993. *Sex, Economy, Freedom, and Community.* Pantheon Books, New York, NY., 208 pp.
7. Ibid.
8. Ralph Waldo Emerson, as quoted in Goodreads.
9. www.goodreads.com/quotes/29365-it-is-one-of-the-beautiful-compensations-of-life-that (accessed January 18, 2022).
10. Margaret Mead (potentially), in her book, *Earth at Omega: Passage to Planetization*, epitaph appearing at the head of Chapter 6. *The Branden Press* (January 1, 1982). See https://quoteinvestigator.com/2017/11/12/change-world/.
11. Chris Higgins. 2008. The Clock of the Long Now. *Mental Floss* (July 29, 2008). www.mentalfloss.com/article/19206/clock-long-now (accessed January 18, 2022).
12. Ibid.
13. (1) The preceding four paragraphs are based on: M. Boyd Wilcox. 1996. *Residents Should Have Right to Cap Population Growth.* Corvallis Gazette-Times, Corvallis, OR. September 10 and (2) John Godfrey Sax. The Blind Men and the Elephant. *The Systems Thinker,* 18. https://thesystemsthinker.com/the-blind-men-and-the-elephant/ (accessed January 12, 2023).
14. (1) Chad E. Cooper. 2018. "We Believe" Statements Are More Important Than Having a Mission Statement! *Thrive Global.* https://medium.com/thrive-global/we-believe-statements-are-more-important-than-having-a-mission-statement-6dc05b14a2e2 (accessed April 5, 2022) and (2) Carrie Melissa Jones. 2020. *How to Create Collaborative Community Mission & Vision Statements.* https://blog.vanillaforums.com/community-how-to-create-collaborative-community-mission-vision-statements (accessed April 5, 2020).

2.6 Discussion Questions

1. How would you define "community"?
2. What does the concept of community mean to you?
3. How would you describe the perfect community?
4. What are some characteristics of a community that is resilient to change?
5. What are some ways we maintain a rich awareness of cultures, histories, and communities, including through our cities' landmark buildings, while implementing the changes required for a more sustainable present and future?

3

Questions We Need to Ask

When Margaret Shannon, a professor of natural resource policy and sociology at the State University of New York, said that "the world does not define itself for us; rather we choose to see some parts of the world and not others,"[1] she opened the door to a whole new way to think about culture: through our individual and collective perceptions. Her statement puts us on notice that we do not clearly *see* our own culture. Rather, we have some perception of it, which in itself creates our culture, because my perception is more or less different from yours, sometimes vastly different.

The amount of conflict in a community is therefore a direct measure of both the difference of opinions and how committed the members are to defend their individual points of view, regardless of their unrecognized narrowness in scope. The purpose of working toward a community vision is to listen to one another, discuss the differences, and move toward a more coherent and harmonious vision for the collective benefit of the community through time, acknowledging each person's perception as part of the community's living culture.

Living culture is thus embodied in the people themselves, and it is there one must search for an understanding of a people as a whole. In this sense, each person is both the creator and the keeper of a unique piece of the cultural tapestry, an understanding of which one can glean only by seeing it simultaneously from many points of view, much as an insect sees. Each perception is composed of many elements, including a holistic understanding of an individual's unique, personal context and the social conditioning leading to his or her cultural foundation. This of course establishes the limits of an individual's understanding.

A person who tends to be positive or optimistic, for example, sees a glass of water as half full, while a person who tends to be negative or pessimistic sees the same glass of water as half empty. Regardless, the level of water is the same. We perceive what our social conditioning dictates, which may have little to do with reality.

The important implication is that the freer we are as individuals to recognize, acknowledge, and change our perceptions without social resistance in the form of ridicule or shame, the freer is a community (the collective of individual perceptions) to adapt to change in a healthy, developmental, and evolutionary way. Each member participating in his or her community must

DOI: 10.1201/9781003353744-3

be respected and have their personal integrity protected by being free to express their feelings, observations, and convictions without fear of ridicule or rejection. These are the ground rules for negotiation.

Before the people of a community are ready to craft a shared vision of their future, they must ask and answer two questions: (1) Who are we today as a culture? (2) What legacy do we want to leave our children?

3.1 Who Are We as a Culture?

Who are we culturally—now, today? This is a difficult but necessary question to contemplate while developing a vision, the nexus between a fading memory of the past and the anticipation of an uncertain future. The members of a community must therefore decide, based on how they define their present cultural identity, what kind of vision to create. A people's self-held concept (based on personal, cultural, and universal values) is critical to their cultural future because their personal and cultural self-image will determine what their community will become socially, which in turn will determine what their children will become socially.

The foregoing question of who we are culturally may be a more important question today than it would have been in the recent past because there are times in history, such as today, when two eras run parallel to each other, when one is dying while the other is struggling with its infancy. This can be a deeply disturbing, confusing, and divisive time as different worldviews, cultural assumptions, patterns, and predominant means of livelihood compete with one another in an effort to give meaning and direction to life.

3.1.1 Experiencing Cultural Disruption

People may feel tempted to retreat into the simplistic solutions often associated with fundamentalism. Fundamentalism (which can ensnare both the political right and left or the spiritual and secular) is characterized by a rigid, impervious belief system that relentlessly widens a perceived polarity between the "safe us" and the "dangerous them." Because it is founded in a fear that feeds on itself, which is always divisive, fundamentalism is not only incapable of tolerating diverse views and backgrounds but also far less capable of creatively asking new questions and discovering new answers within a context of dynamic complexity—effectively locking the community in a dysfunctional cycle, unable to move ahead.

In any age regressive, fundamentalism is simply not up to the challenge of our times. Instead, the next stage of cultural evolution must focus inward, into each person's mindful self-awareness—the only realm out of which can develop creative, self-organizing innovations that offer sustainable ways of living based on the quality of both interpersonal relationships and those between humanity and its environment.

Like all evolution, cultural evolution thrives in a context rich in diversity and complexity wherein myriad opportunities for interaction exist. Self-organizing innovations can emerge out of such a setting as people search for ways to live consciously and sustainably in every sense of the word. These innovations become "attractors," which draw us out of the chaotic, negative cycles into further experimentation with social-environmental well-being.

The most powerful attractors are those that respond to people's basic requirements for survival and to their deepest yearnings for connection, meaning, and transcendence—the ingredients of personal wholeness and self-actualization. When these attractors resonate among large numbers of people (a critical mass), society shifts, but people must first be aware of these "attractors" amid the flotsam and jetsam of change in which the decay of the dying era seems, at least momentarily, to overwhelm the formative one.

Of course, there initially is a multitude who, preferring "the devil they know to the devil they don't," steadfastly swear allegiance to the passing era (or to a nonexistent "golden age") by clinging tenaciously to old views and ways of doing things. In counterpoint, there is an expanding, energetic group of younger people who find the present and future ripe with possibilities. And it is here, in the present, that small choices and actions can have major, albeit unpredictable, effects in determining what comes next and how it manifests.

And somewhere among the millions of choices and thousands of experiments with conscious living is the possibility that they will coalesce into a new society that is founded on and protects true community while endowing life with real meaning. For such a civilization to be viable, however, it would have to be anchored in the bedrock value of social-environmental sustainability in all its various aspects. It will never be perfect nor an imagined "golden age," but it will be the best we have to further design and build on.

There is an old adage (misattributed to Thomas Jefferson): "In matters of taste, swim with the current. In matters of principle, stand like a rock."[2] To identify those principles and/or values on which we stand firm, we can ask the following: What are the fundamental principles that I believe in to the point of no compromise? What values are central to my being?

3.1.2 Categories of Value

When we as individuals clearly understand and can explicitly articulate our personal values, then we can live accordingly. Let us consider three categories of values: universal, cultural, and individual.[3]

- Universal (or archetypal) values reveal to us the human condition and inform us of our place therein. Through universal values, we connect our individual experiences with the rest of humanity (the collective unconscious) and the cosmos. Here, the barriers of time and place and of language and culture disappear in the ever-changing dance of life. Universal values must be experienced; they cannot be comprehended. Can you, for example, know a sunset? Fathom a drop of water? Translate a smile? Define love?

Universal values are the timeless constants emerging in different cultures at various times throughout history. "Even as the hands of a clock are powered from the center [post], which remains ever still, so the universal values remain ever at the center of human life, no matter where the hands of time are pointing—past, present, or future."[4] These are some of the truths comprising the human condition toward which people aspire (such as joy, unity, love, and peace); of these, the sages have spoken in every language, every age.

- Cultural values are those of the day and may be generally agreed upon. They are established to create and maintain social order in a particular time and place and can be highly volatile. Cultural values concern ethics and human notions of responsibility to one another, the community, and the planet; notions of right and wrong and of good or evil; and norms, customs, and manners.

  Culture reflects the ideas and behaviors a society rewards or suppresses according to its perceived alignment with accepted values. Hence, cultural values are for an individual a mixed bag, especially in highly complex contemporary developed societies where previous bedrock principles may have eroded: coherence and stability of the family, community, and a common mythology.

  Every culture is a person in a sense, and like people, there is simultaneously the potential for creative and nurturing interaction and competition or conflict when cultures meet. Although we are all too familiar with cultural conflicts and their destructive capacity, it is well to remember that a meeting of cultures also triggers tremendous explosions of creativity in such things spanning food, art, architecture, medicine, language, ethics, education, law, philosophy, and governance.

- Individual values are constituted by the private meanings we bestow on those concepts and experiences (such as marriage vows or spiritual teachings) that are important to us personally. These meanings spring from how we are raised by our families of origin and which of our parents' values we take with us as formative. These influences may become shaped by our own exposure to new ideas, our life experiences, and how much we are willing to grow and change psychologically and spiritually as a result. As such, individual values are reflected in such things as personal goals, humor, relationships, and commitments.

Thus, how well a people's core values are encompassed in a vision depends first on how well people understand themselves individually and as a culture, which means how well they understand their core values, and second on how well that understanding is reflected on paper, where there can be no question about what has been stated and how. By way of illustration, let us consider a Canadian First Nation, as Canada refers to its Indigenous Canadians.

Some of the Canadian First Nations have departed from their ancestral culture because they have, against their will, been forced to adopt European-Canadian ways, which means they have given up or lost much of their traditional lifestyle. Yet they have not, by choice, totally adopted white values and want to retain some degree of their early culture. Thus, the three questions they must ask and answer are as follows: Which of our ancestral ways still have sufficient cultural value for us to keep them? Which of the white ways do we want to or are we willing to adopt? How do we put these chosen elements together in such a way that can today define who we are culturally?

In 1993, I (Chris) was asked to review a document, written by a timber company, for a First Nation in western British Columbia, Canada, whose reservation is located between the sea and land immediately downslope from that which a timber company wanted to cut. The problem stemmed from the fact that the company could only reach the timber it wanted by obtaining an easement through the reservation, which gave the First Nation some control over the proposal. The First Nation wanted to have an active voice in how the company would log the upper-slope forest because the outcome would affect the reservation and its water catchments for many years.

By virtue of the company's required easement through the First Nation's land, the First Nation's requirements were the strong organizing context that would shape the behavior of the company as it logged the upper-slope forest. If, however, the company had not been required to pass through the First Nation's land, the issue could, through conventional logging practices of the time, have led to a series of cumulative impacts that could have destroyed the First Nation's land and inherent cultural values relating to their land for generations.

Before meeting with the timber company, the First Nation's chief asked me for some counsel. My reply was as follows:

> Before I discuss the document I've been asked to review, there are three points that must be considered if what I say is to have any value to the First Nation. What I'm about to say may be difficult to hear, but I say it with the utmost respect.
>
> *Point 1:* Who are you, the First Nation, in a cultural sense? You are not your old culture because you have—against your will—been forced to adopt some white ways, which means you have given up or lost ancestral ways. You are not—by choice—white, so you may wish to retain some of your ancestral ways. The questions you must ask and

answer are as follows: (1) What of our ancestral ways still have sufficient value that we want to keep them? (2) What of the white ways do we want to or are we willing to adopt? (3) How do we put the chosen elements of both cultures together in such a way that we can define who we are as a culture today?

Point 2: What do you want your children to have as a legacy from your decisions and your negotiations with the timber company? Whatever you decide is what you are committing your children, their children, and their children's children to pay as the effects of your decisions unto the seventh generation and beyond. This, of course, is solely your choice and that is as it should be. I make no judgments. But whatever you choose will partly answer Point 3.

Point 3: What do you want your reservation to look like and act like during and after logging by the timber company? How you define yourselves culturally, what choices you make for your children, and the conscious decisions you make about the condition of your land will determine what you end up with. In all of these things, the choice is yours. The consequences belong to you, your children, and beyond.

Communities everywhere can similarly collaborate on a quest for sharing observations, experiences, ideas, meaningful discussion, and a transparent, creative process to shape their next and ongoing vision. Such a keystone process has a life of its own; the process requires an investment of time and sincere intention and attention. What about you, the reader? Who are you today? We each change personally as we grow in years and are shaped by our experience, as our respective communities also change. Each community that wishes to create a vision for a sustainable future must therefore ask itself this: who are we today in a cultural sense? Based on how a community sees itself, each community must then ask this: who do we want to be or to become in the future? These are important questions and must be clearly answered on paper for all to see because how they are answered will determine the nonnegotiable constraints that set the overall direction of a community's vision and thus the legacy its new generations will inherit. To answer who we are as a community today and what we want as a community in the future, it is advantageous to begin by honestly evaluating your own personal set of values.

3.1.3 Identifying the Values You Wish to Pass On to the Future

Although it may not seem important at any particular moment in a given day, it is critical in the long run to know what values to safeguard in one's community. After all, values shape the contours of our lives. For example, a simple act by the very people who went to Phoenix, Arizona, to find relief from their allergies has placed Arizona among the top 10% of states in pollen count during the six-week allergy season.

Before urban sprawl began consuming the desert, the area around Phoenix was a haven for people who suffered from allergies. Doctors in the 1940s and 1950s sent patients there because the dry air was virtually pollen-free. But many of those people also brought with them their non-desert plants, which subsequently matured and now fill the air with pollen during the spring of each year.

In addition, the dry climate causes pollen grains from non-indigenous plants to stay aloft and ride the air currents, wafting in every zephyr. They are not washed from dry desert air, as they are in non-desert areas that experience spring rains. So the allergy sufferers themselves made their own haven into their worst nightmare by not identifying and protecting the very value that brought them to Phoenix in the first place—air virtually free of pollen.[5]

Mikhail Gorbachev, in an interview with Fred Matser, expressed concern that "we are witnessing a breakdown of the proper relationship between human kind and the rest of nature. I believe . . . this situation has arisen," says Gorbachev, "because we have retreated from the perennial values. I don't think that we need any new values. The most important thing is to . . . revive the universally known values from which we have retreated."[6]

With this in mind, you would be wise to pause and describe to yourself how you feel about your community before you begin to craft your vision and goals. What types of images come to mind? Who do you think about in your community and why? What places do you think about (open space, such as parks and playgrounds, bike paths, community gardens, or shopping malls and schools)? Do activities present themselves? If so, describe them. In short, characterize your community, and be sure to do so either by recording your questions and answers orally or in writing.

If you find that you are unsure how you feel about something, take the time to consciously observe your community; see how it functions and how you feel about the way it functions. How friendly is it? How safe do you feel living there, and moving about during the day and at night?

As a thought experiment to reveal more information, put yourself in the position of a consultant who has been hired to characterize your community. What questions would you, as a consultant, ask the residents? Why did you select these particular questions? What are you hoping they will tell you? Why do you think these particular questions are important? Now continue your observations and answer the questions for yourself.

Based on what you see and feel, what values do you hold that are met in your community and why? Which values are not met and why? By asking, it becomes clear that precisely framing, insightful questions are the key to crafting a good vision statement and goals.

Now, using this technique, characterize and design the community of your dreams. What would it be like? Can you see where, how, and why your interests and talents would fit into your vision? Describe in writing its primary elements, remembering that the most important part of community, by the very nature of the concept, revolves around the quality of human relationships and the reciprocal partnership between the community, its neighborhoods, and its adjacent landscapes.

If even a small group of community members is willing to participate in such a personalized exercise, it would quickly become apparent that the makings of a sound vision and goals are contained in the collection of personal observations, feelings, and values. But taken alone, personal values are not enough. The pulse of the community as a whole must be taken.

3.1.4 Ferreting Out Community Values

To ask a relevant question about where you are going, you must not only know where you want to go but also where you are, which means taking stock of who you are. A shared vision is a statement of where you want your community to go. It follows that assessing your community, including the present reciprocity of its relationship to the immediate landscape, provides your starting point for the journey.

We often come together in one location to hold a town meeting to discuss one or more initiatives or ideas. However, going out into the community can be extremely important because it allows us to gain a truer perspective. Looking at the community from multiple vantage points is important. One way of assessing a community is by entering into and directly experiencing its routines. This means selecting people to attend school events, such as football games and meetings of the parent-teacher association; visiting people in their kitchens and living rooms; and going into cafes, gas stations, laundromats, and other places where people gather, such as taverns and churches. The purpose of these visits is to interact with residents to determine such things as what they do for work and what their work routines are, what their personal interests and recreational patterns are, what support services are important to them, and how they feel about changes within the community and between the community and its landscape.[7]

To really understand how a community sees itself, one must ask people not only what they like about their community and its landscape and why but also what they do not like and why. One must ask people what they would most like to change about their community and its landscape. Questions also help one find out which informal networks people use both to communicate with one another and to solve problems, as well as who they trust and rely on as communicators and caretakers.

Presently, groups use social media to identify social networks. A community visioning project can—and must—strongly consider having an interactive social media site to exchange ideas and provide readings, photos, oral histories, and case studies from similar communities to ensure that everyone is able to exchange information toward the development of the community's vision. This is a standard way to understand networks for outreach, education, and marketing (which is not limited to commercialism). If a community visioning process entertains engaging a consultant, likely that organization's work plan could set up and manage a social media tool for a community who wished to use it.

Alternatively, communities may choose to hire a consultant to design survey questions and derive the answers by visiting personally and informally with community members in both their places of business and their homes. It is imperative that the community members and their consultants establish and maintain trust, which *must* include neutrality and confidentiality. People do not care how much a person knows until they know how much that person genuinely cares about them and their project.

It is, after all, the quality and sustainability of one's unique community that are being mapped into the future, and that is no small matter. It is thus important to understand that trust is heightened and the community's purpose is served to the extent that members of the community become actively engaged in the process. Building and maintaining relationships of trust take time and do not occur overnight.

The purpose of asking the questions outlined earlier is to make the informal system of community clearly visible in such a way that, by understanding the range of members' concerns and how they see themselves in relationship to those issues, one can help the community recognize and express its current cultural identity. This kind of information is called ethnography, meaning "the story of the people."

The story of the people as a baseline description of how community residents identify culturally is a sound preparatory step toward crafting a shared vision. A sustained process of interaction within a community at the informal level has two important effects:

1. It fosters involvement, empowerment, and the community's self-determination. Personal and social reflection not only determines the intelligence and possible consequences of any given action but also leads citizens to see what the next step might be and to take it. It is thus important, as French philosopher Henri Bergson observed, to "think like a man [person] of action, and act like a man [person] of thought."[8]
2. It can prompt social institutions into becoming more responsive because people within agencies gain insight into the concerns of citizens and thus into a community's cultural identity by participating in the ongoing "story of the people." Such participation gives agency people good and relevant information that makes sense to the citizens and allows them to understand why citizens say what they do.

Public opinion is characterized as much by emotions as by logic and is determined by self-interest, which can be expanded from a strictly individualistic self-interest into an enlightened (more conscious), broader (community) self-interest.

One important fact must be kept in mind, however; information alone will rarely change the ideas and opinions held by people, especially if the information is in a purely abstract form. Good public relations work more on the "hidden" levels of thought and feelings, those exemplified by the trust embodied in genuine, mutual goodwill, as opposed to the level of logic that is purely intellectual.

For example, Mahatma Gandhi, by insisting on meeting the downtrodden masses of India on their own ground, established the credibility and genuineness of his understanding of their plight and of his love for them, which made all the difference because they felt not only that they knew him but also that they could trust him. And trust is the key to one's belief in and willingness to follow the lead of another person.

One of the ways Gandhi garnered the people's trust was to deal fairly with all sides (including the British, regardless of their behavior), which is critical because almost any subject poses controversy when necessarily viewed from more than one perspective. This means that one must carefully guard against arrogance in one's own point of view and behavior because arrogance will almost always cause the listeners to resent whatever is being said, not matter how well founded it may be, and thus to stop listening.

Consider that the intangible asset of "goodwill," analogous to trust, is rated surprisingly high in value as part of the total price of purchasing a business. Building the intangible asset of goodwill takes many years of diligent, consistent effort and communication. And, once earned, few organizations are willing to relinquish this asset. Individuals, especially leaders, must also mindfully build this asset over time in order to be effective.

For example, someone trying to lead a community toward social-environmental sustainability must make it his or her business to supply accurate information and other useful services to those citizens involved in the visioning effort. It would be folly to try to mislead them with exaggerated, overly optimistic statements or facts of dubious merit. In return for such steadfast honesty, fellow citizens listen to and carefully consider what a leader says, even though they may not agree.

Earning the trust and goodwill of one's fellow citizens is even more critical today. Some community members may feel that they have been lied to by leaders or experts in the past. People may have good reasons to be skeptical and thus reluctant to place their trust in someone they do not personally know. In any event, they may be cautious. Nothing persuades an audience to examine your point of view as much as personal trust and clear, direct communication. The conclusion in your presentation is only effective if the audience senses authenticity, sincerity, and beneficial options and opportunities supporting the consensus and realistic pathways to get there.

People are primarily motivated by their own self-interest. Any method used to gather information within the community must be based on personal trust and goodwill. But it is the people who must ultimately craft their vision and goals with the help of a neutral, third-party facilitator. Experience shows that a persuasive message is more likely to be accepted and acted upon if it meets the following criteria: (1) it provides for a personal necessity or desire, (2) it is in harmony with group beliefs, and (3) the audience is led to the final conclusion and then left to discover it for themselves.

Through the process, an initial, draft-proposal document must be produced on which the community can comment during a period long enough to give people adequate time to deeply consider the contents. And then the people must really be listened to and their comments collected, collated, and incorporated into the vision. A vision, to be effective, must be finalized by consensus. And finally, a vision and its goals must periodically be reexamined (re-envisioned) to keep it dynamic and relevant to the present and future, a critical step that is seldom taken.

Such small-scale change, done with and by people rather than for or to them, when multiplied over a whole community, becomes a clear signpost toward a community's vision of social-environmental sustainability and hence the legacy people leave their children.

3.2 What Legacy Do We Want to Leave Our Children?

Once a group of people, whether a community, such as the Indigenous Canadians, or your own hometown, has defined itself culturally (present and future), it can decide what legacy it wants to leave its children. This must be done consciously, however, because the outcome or consequences of whatever decisions the group makes under its reassessed cultural identity is what the group is committing its children, their children, and their children's children to live with.

The rest of my reply to the First Nation in Canada:

> This is a difficult task at best because, like any definition (community, tribe, culture, or reservation), it is a human invention and has no meaning to nature. Therefore, you must tell the timber company, clearly and concisely, what the terms in this ecological brief mean to you and how you interpret them with respect to the company's actions that will affect your reservation.
>
> 1. Every ecosystem functions fully within the limits (constraints) imposed on it by nature and/or humans. Therefore, it is the type, scale, and duration of the alterations to the system—the imposed limitations—that you need to be concerned with.
>
>   If your reservation looks the way you want it to and functions the way you want it to, then the question becomes this: how must we and the timber company behave to keep it looking and functioning the way it is? If, on the other hand, your reservation does

not look the way you want it to and does not function the way you want it to, then the question becomes this: how must we and the timber company behave to make it look and function the way we want it to?

But regardless of your decisions or the company's actions, your reservation will always function to its greatest capacity under the circumstances (constraints) nature, you, and the company impose on it. The point is that your decisions and the company's actions, excluding what nature may do, will determine how your reservation both looks and functions. This reflects the importance of your answer to this question: what do you want your reservation to look like and how do you want it to function *after* the timber company has left? It also reflects the importance of what you decide.

2. If you want the landscape of your reservation to look and function in a certain way, then how must the timber company's landscape look and function to help make your reservation be what you want it to be? Keep in mind that the landscape of your reservation and the company's timber holdings are both made up of the collective performance of individual stands of trees or "habitat patches" (a "stand" is a human-delineated group of standing trees). Therefore, how the stands look and function will determine how the collective landscape looks and functions.
3. Remember that any undesirable ecological effects are also undesirable economic effects over time. Your interest in your reservation will be there for many, many years, generations perhaps, but the company's interest in the forest may well disappear just as soon as the trees are cut. So the company's short-term economic decision may be immediately good for them but may, at the same time, be both a long-term, negative ecological and economic decision for you.
4. To maintain ecological functions means that you must maintain the characteristics of the ecosystem in such a way that its processes are sustainable. The characteristics you must be concerned about are (1) composition, (2) structure, (3) function, and (4) nature's disturbance regimes, which periodically alter the ecosystem's composition, structure, and function.

   The composition or kinds of plants and their age classes within a plant community create a certain structure that is characteristic of the plant community at any given age. It is the structure of the plant community that in turn creates and maintains certain functions. In addition, it is the composition, structure, and function of a plant community that determine what kinds of animals can live there, how many, and for how long. If you change the composition, you change the structure, hence the function, and you affect the animals. People and nature are continually changing a community's structure by altering its composition, which in turn affects how it functions.

For example, the timber company wants to change the forest's structure by cutting the trees, which in turn will change the plant community's composition, which in turn will change how the community functions, which in turn will change the kinds and numbers of animals that can live there. These are the key elements with which you must be concerned because an effect on one area can—and usually does—affect the entire landscape.

Composition, structure, and function go together to create and maintain ecological processes both in time and across space, and it is the sustainability of the processes that, in the end, creates the forest. Your forest is a complex, living being, not just a collection of commercially valuable trees—as the timber industry usually thinks of it.

5. Scale is an often-forgotten component of healthy forests and landscapes. The treatment of every stand of timber is critically important to the health of the whole landscape, which is a collection of interrelated stands.

   Thus, when you deal only with a stand, you are ignoring the relationship of that particular stand to other stands, to the rest of the drainage, and to the landscape. It's like a jigsaw puzzle, where each piece is a stand. The relationship of certain pieces (stands) makes a picture (drainage). The relationship of the pictures (drainages) makes a whole puzzle (landscape). Thus, the relationships of all the stands within a particular area make a drainage, and the relationships of all the drainages within a particular area make the landscape.

   If one piece is left out of the puzzle, it is not complete. If one critical piece is missing, it may be very difficult to figure out what the picture is. So each piece (stand) is critically important in its relationship to the completion of the whole puzzle (landscape). Therefore, the way each stand is defined and treated by the timber company is critically important to how the landscape, encompassing both the company's land and your reservation, looks and functions over time.

6. Degrading an ecosystem is a human concept based on human values and has nothing to do with nature. Nature places no extrinsic value on anything. Everything just is, and in its being it is perfect (intrinsic value). Therefore, when considering intrinsic value, if something in nature changes, it simply changes—no value is either added or subtracted. But superimposing the extrinsic value of human desires on nature's intrinsic value creates a different proposition. Thus, whether or not your reservation becomes degraded depends on what you want it to be like, what value or values you have placed on its

being in a certain condition, to produce certain things for you. If your desired condition is negatively affected by the company's actions, then your reservation becomes degraded. If your desired condition is positively affected by the company's actions, then your reservation is improved. Remember, your own actions can also degrade or improve your reservation.

7. It is important that you know—as clearly as possible—what the definitions in this brief really mean to you and your choices for your children and your reservation. Only when you fully understand what these definitions mean to you can you negotiate successfully with the timber company.

If you, the reader, substitute the name of your own community wherever "reservation" occurs in the brief, and if you substitute "proposed development," "land-use zoning," or "land-use planning" wherever "timber company" occurs in the brief, you have an outline to follow for your own community, perhaps with a few modifications to fit specific local conditions.

In addition, we must exercise the good sense and humility to sincerely *ask* our children (beginning with second- and third-graders) what they think and how they feel about their future, as we build our shared visions of a sustainable future in which each person's core values (including our children's) are acknowledged.

3.3 Including Children in Envisioning the World They'll Inherit

Consider for a moment that children must inherit the world and its environment as we adults leave it to them. Our choices, our generosity or greed, our morality or licentiousness will determine the circumstances that will become their irreversible reality.

Why, then, do adults assume they know what is best for the children and their future progeny when most adults are degrading their world through short-term decisions, fear, greed, and competition? Why are children seldom if ever asked what they want us to leave them in terms of true social-environmental quality? Why are they never asked what kinds of choices they would like to be able to make when they grow up?

Similarly, we might initiate conversations with our own elders regarding where we have come from based on their legacy to us. Then we might consider what we have lost in terms of social-environmental sustainability due to the outcomes of their own or their generations' decisions and actions and how we seem to repeat the same mistakes.

If we do not know what is good for us, it is disingenuous to try to speak for children and future generations. This lack of responsible care was keenly explored by the global community at the June 1992 United Nations Conference on the Environment and Development (UNCED), held in Rio de Janeiro (cite to: https://www.un.org/en/conferences/environment/rio1992). A 12-year-old girl delivered to the entire delegation a most poignant speech about a child's perspective of the adult's environmental trusteeship. I (Chris) saw a video of the speech in which a child was pleading for a more-gentle hand on the environment so that there would be some things of value left for the children of the future. The adult audience was moved to tears—but not to action.

If our human society is to have a sustainable future, it is increasingly important to listen to what the children say because they represent that which is to come. Children have a beginner's mind. To them, all things are possible until adults with narrow minds, who have forgotten how to dream, put fences around their imagination. Too often we adults think we know what the answers should be yet can no longer see what they might be. To us—whose imaginations were stifled by parents, schools, and frightened peers—things have rigid limitations. We would do well, therefore, to consider carefully not only what the children say is possible but also what they want.

In asking children what kind of world they want us to leave them, we must understand how best to talk the matter over with them on their level. The same basic principles that apply to speaking with adults apply to speaking with children. The difference is changing the emphasis and intensity to fit the children's age and circumstances.

In speaking to adults, for example, one makes eye contact, uses visuals, encourages participation by the audience, and varies one's vocal delivery and gestures. People who tell stories to children or employ other ways to educate them use the same techniques, only in an exaggerated manner, and the younger the children, the greater the exaggeration. One must also make sure that the children are interested in what is being said and that the topic, such as what they want their future to be like, is presented in such a way that it is important to them, and they understand how and why that is so.

3.3.1 Encourage Children to Draw Their Visions of Community

The younger the children, the more they enjoy the here and now, and the less they relate to the notion of past and/or future tense; after all, to them, yesterday was a long time ago and tomorrow is hard to grasp. Older children may more readily relate to past, present, and future events. As with adults, communication is much clearer and more meaningful for them if the ideas under discussion are connected to something they already know.

Here are some specific points to consider:

1. *Arrange the room for optimum attention* by asking yourself the following: Will everyone be able to see? (2) Will they be comfortable? (3) Will each child feel like she or he is a part of the group? As it turns

out, the best seating arrangement for a young audience is a semicircle because the children can interact more easily with the speaker. Making them part of a semicircle is the first step in helping them feel like they are really involved in the discussion. Children tend to lose interest if they do not feel really involved with the group. In my (Holly's) experience, this format is best with small groups of adults as well.

It is best if the speaker can sit down close to the children because it is less threatening for them to be at eye level rather than having to peer upward at a looming giant of an authority figure. Some people have the best results when everyone, including the speaker, sits on the floor. This arrangement is also true of adults—everyone on an even level.

2. *Involve the children:* Although young children can be asked to act out their feelings, middle school and high school kids may be too self-conscious to participate in such a manner because they do not want to appear foolish in front of their peers. With older children, ask a question and have them think about it or raise their hands in response. Ask questions in the beginning that everyone can answer positively and then incorporate the information into the discussion.
3. *Know the audience:* Before speaking to a group of children, do a little fact-finding. What have they been studying in school that relates to what they want their future to be like? Then use this material to draw and hold their attention.

   Ask the children about themselves: (1) What do you like best or are you most interested in? (2) Do you know what you want to be when you grow up? (3) How do you feel about the world around you? (4) What do you want the world to be like when you are grown up? (5) What do you want the world to be like for your children? Although these questions are vital because finding out about the kids helps you connect with them, it is even more important for the kids to know that you genuinely care about them and their future. It is thus important to relate your points to what they are thinking about.

   You must also adjust your presentation to the age group you are visiting with. For example, exaggerated gestures are okay with young children but must be toned down for older ones. And while young children can deal with open-ended questions ("What do you think?"), older kids are often too embarrassed to say what they think in front of their peers. In this case, an open-ended question inhibits participation.
4. *Variety:* Everyone responds to variety, especially children. Variety can be created with a story (including how you project your voice, which is important in holding kids' attention), listening and learning (presentation) speaking and sharing (questions and answers,

including a list of questions of your own gleaned from things the children say to you and/or one another), and open participation (discussion), all of which work best if appropriately laced with drama and humor. In addition, visual aids (such as a flip chart, showing a film or slides) help to emphasize or clarify your point, and props (such as giving a demonstration, putting on a skit, reciting a poem, or even giving a short assignment) promote involvement. Of course, any combination of these can be used.

Pictures are also a good way to get to know children, especially young children, and to incorporate variety into one's discussion of what they want us, as adults, to leave for them as a legacy. My next-door neighbor, Justin Lewis, taught a combined class of second- and third-graders. Each year I spoke to his class about some aspect of nature, usually forests. Together, we developed a way of helping young children express what they want the forests to be like when they grew up—in other words, what legacy they want us to leave for them.

Justin began the process, before I came into his class, by asking the children to draw individual pictures of what they thought a forest was, and he had them write an explanation to accompany the picture. He gave me the pictures before I visited with the children so I could evaluate their perception of a forest.

I then visited with them in their classroom and showed them slides of a forest with its plants and animals. We talked about plant and animal relationships within the forest (including people). Next, we took a field trip into a real forest to help the children transfer any abstract ideas into concrete experiences to further their understanding.

Once again, Justin had them draw individual pictures of a forest so they could incorporate their new understanding. Justin and I then discussed the difference between their first and second pictures. Finally, so the children could begin to understand that it was everyone's responsibility to save, protect, and care for the forests for the future, which was not always easy, Justin had them work collectively to draw a mural of the forest on a chalkboard—the forest they wanted us, the adults of the world, to leave for them. The mural was important because they began learning to work together toward a common goal, the precursor to a full-fledged vision.

5. *Keep it short:* Time is a critical factor when visiting with children because a child's attention span is much shorter than that of an adult. For younger elementary school children, 10 to 15 minutes generally works well, while older children will usually pay attention for 30 to 50 minutes.
6. *Make sure the children understand you:* The language and concepts you use must be kept simple for small children, which by no means precludes teaching them some new things, such as words or ideas. Be

sure to check with the kids when using words that might be unfamiliar. Ask, "Can anyone tell me what this is (or means)?" Make sure everyone understands before moving on.[9]

If you are in doubt about how to make certain that children of different age groups will understand you, take the time to visit teachers and librarians. They have a wealth of tips on different ages, developmental stages, and interests to help you ask the children what they want us, as their guardians, to protect for them as our legacy to their livable future.

The future, after all, is theirs, which brings us to the necessary topic of vision, goals, and objectives. For your community to negotiate effectively for the future, for yourself, for your children, and for their children, your community must be clear about its vision, goals, and objectives.

3.4 References

1. Margaret Shannon, in Chris Maser. 1999. *Ecological Diversity in Sustainable Development: The Vital and Forgotten Dimension*. Lewis Publishers, Boca Raton, FL., 402 pp.
2. The Thomas Jefferson Foundation's Monticello Organization (Jefferson Encyclopedia) indicates this is a spurious quotation, first attributed to Jefferson around 1973, although one version of the quote seems to have first appeared around 1891. www.monticello.org/research-education/thomas-jefferson-encyclopedia/matters-style-swim-currentspurious-quotation/.
3. Laurence G. Boldt. 1999. *Zen and the Art of Making a Living*. Penguin Books, New York, NY., 608 pp.
4. Ibid.
5. The Associated Press. 1997. *Ah-choo, Arizona no Longer Haven for Allergy Sufferers*. Corvallis Gazette-Times, Corvallis, OR., March 25.
6. Fred Matser. 1997. Nature is My God. *Resurgence*, 184:14–15.
7. Kevin Preister. 1966. Community Assessment. *Community Ecology, A Newsletter of the Rogue Institute for Ecology and Economy*, 2:1, 5.
8. Henri Bergson. 1950. *The Forbes Scrapbook of Thoughts on the Business of Life*, p. 442.
9. The foregoing discussion on communicating with children is based on the following: (1) Patricia L. Fry. 1997. Speaking to Kids. *The Toastmaster*, 63(5):24–27; (2) Chapter 9—Ask The Children—in Chris Maser. 2004. *The Perpetual Consequences of Fear and Violence: Rethinking the Future*. Maisonneuve Press, Washington, DC., 373 pp; and (3) my (Chris') experience in classrooms with third-, fourth-, and fifth-grade students, where I helped to teach them about nature, as well as other age groups.

3.5 Discussion Questions

1. What are some ways we can achieve social-environmental sustainability, equity, and inclusion in an ever-more-diverse society?
2. How can we build and maintain safe, sustainable, and thriving communities in a culturally and politically diverse society?
3. Are there core human values that must be maintained in a community if it is to be socially sustainable with dignity for all? What values would be on your list and why?
4. What kind of social-environmental legacy would you want your community to pass forward to its children that might encourage them to engage in and remain part of the community?
5. Why is it important to include children in a community's decision-making processes?

4

Understanding a Vision, Goals, and Objectives

4.1 Introduction

Having defined who they are culturally and having determined what legacy to leave their children and the kind of legacy their children want them to leave, the people of a community are now ready to craft a vision of what they want as a sustainable future.

Although the word vision is variously construed, it is used here as a strong, organizing context in the form of a shared view of the future based on three overlapping aspects: worldview, perception, and imagination.[1]

4.2 Worldview

Our worldview is our personal understanding of how the world works, our overall perspective from which we interpret the world and our place in it. However, it can also be seen as a metaphysical window to the world, which cannot be accounted for on the basis of empirical evidence any more than it can be proved or disproved by the argument of fact. "Metaphysical" comes from the Greek *meta*, "beyond," and *physic*, "physical, nature."

Two ubiquitous worldviews concern the sacred and the commodity. One need not be religious in the conventional sense to hold a sacred view of life, because a sacred view focuses on the intrinsic value of all life. "Sacred" comes from the Latin *sacer*, "holy." A sacred view of life corresponds to the Sanskrit *sat*, *cit*, and *ananda*, or "being," "consciousness," and "bliss." As such, it gives birth to feelings of duty, protection, and love while emphasizing the values of joy, beauty, and caring or lovingkindness. A sacred view of life suggests holding an ethic of self-control to destructive behavior against others and nature.

DOI: 10.1201/9781003353744-4

In contrast, a commodity view of life concerns control, management, and profit by exploiting nature for economic gain. At the core of the commodity worldview are several 18th- through 20th-century economic seeds, such as self-interest, the delusion of an ever-growing economy in the self-serving present stemming from "rational economic man," and others.[2] It is necessary, with respect to the commodity worldview, to protect the biophysical processes of social-environmental sustainability for the benefit of *all generations* through enacting *legal* constraints on the over-exploitive, myopic human behavior entrenched in the money-making mindset.

Eighteenth-century British philosopher and statesman Edmund Burke (1729–1797), considered a founder of modern conservatism, understood well the need for external constraints on the destructive appetites of humanity when he penned:

> Men [people] are qualified for civil liberty in exact proportion to their disposition to put moral chains upon their own appetites. . . . Society cannot exist unless a controlling power upon will and appetite be placed somewhere, and the less of it there is within, the more there must be without. It is ordained in the eternal constitution of things that men [people] of intemperate minds cannot be free. Their passions forge their fetters.[3]

If, therefore, we are going to refresh and align our worldview to live within our planetary boundaries[4] we must begin by focusing on assumptions and attitudes, not facts alone. As Albert Einstein said, "The world we have made is a result of . . . [the] level of thinking we have done thus far, creates problems we cannot solve at the same level at which we created them." In essence, an outer change always begins with an inner shift, which Einstein referred to as, "a new level of thinking."[5] In essence, we must raise the level of our social-environmental consciousness to reflect the wisdom of our perceptions, choices, and decisions and their outcomes for the sake of all generations.

4.2.1 Perception

Our perception is the vision with which we see the world and interpret what we see based on our own experience. Our perception colors our future experiences—this is human. "Perception" comes from "perceive," which is from the Latin *percipere*, "to seize wholly."

To see wholly is to see in a better way; therefore, to perceive a problem clearly means to look at it from every angle, every perspective in order to begin to relate its parts and history, formulating options toward its solution. To solve a problem or resolve a conflict, we need the wisdom to keep searching and the confidence to hold what we find up to the light. It is when we doubt our capacity to love and to create, which we then replace with fear and isolation, that we begin to further distort our worldly perceptions. To quote writer Suzy Kassem, "Doubt kills more dreams than failure ever will."[6]

When our view of the world is based on love and the confidence of clear perception, we become members of the global community leading the world to a greater understanding.

Nevertheless, we are usually moved from dread, avoidance, and confusion to attention and clarity only when we perceive the necessity to do so. "Necessity," wrote Plato, "is the mother of invention."[7] Necessity, in this sense, is simply the perception that a current situation has become intolerable and that something must be done about it, which means the perceiver is the one who must act.

Once necessity is acknowledged and *accepted*, we begin searching for a solution to our problem by examining our assumptions and perceptions, which jogs us out of our current prejudices and conceptual limitations in such a way that we can sift through those old ideas and concepts that we have in the past overlooked, disregarded, and/or discarded. Fortunately, the urgency we feel when we acknowledge necessity can help us move forward instead of too easily giving up searching for and practicing a better way of being and doing.

These notions call forward our individual and collective choices, namely, to yield to the comfortable blindness of ignorance or to summon our courage and make resolute our determination to search until we find a better way. It has been wisely said that anything will reveal its secrets if you love it enough.

4.2.2 Imagination

Imagination, or seeing that which can be, is the third aspect of vision. Even as we open our physical eyes and see the world as we think it is, with all its problems and opportunities, we can open the mind's eye to reveal the possibilities of unseen future realities. To open our physical eyes fully, we must learn to trust so we can accept "what is, as it is" through the eyes of love.[8] To open the eye of our mind, we must learn to trust that what we see in our imagination we can bring forth in the physical world.

Whereas perception involves seeing that which already exists in the outer world, imagination involves seeing in one's inner world that possibilities can be made manifest in one's outer world. Albert Einstein is sometimes credited with saying: "Your imagination is your preview of life's coming attractions,"[9] to which William Butler Yeats added, "In dreams begins responsibility."[10] Consider, therefore, that everything humanity has ever created (or ever will create), both tangible and intangible, began as a single idea in the privacy of someone's mind, be it this book, a religious order, or going to the moon. Our imagination is the source of our power to create and the driving force behind our choices—the prerequisites of a shared vision, goal, or objective toward which to build. Imagination is the heart of empathy, and empathy opens pathways not otherwise possible.

4.3 Defining Vision, Goals, and Objectives

Alas, few people know what a vision, goal, or objective is; how to create them; how to state them; or how to use them as guidelines for sustainable development.

4.3.1 Vision

Defining a vision and committing it to paper goes against our training because it must be stated as a positive in the positive, something we are not used to doing. Stating a positive in the positive means stating what we mean directly. For example, a local community has an urban growth boundary that it wants to keep within certain limits, which can be stated in one of two ways: (1) we want our urban growth boundary to remain within half a mile from where it is now situated (a positive stated as a positive), or (2) we don't want our urban growth boundary to look like that of our neighbor (a negative that one is attempting to state as a positive). It is more motivating and memorable to state a preference or a goal in the positive.

Pertinent adages include the only thing that is permanent is change; consequently, there is no growth without risk. Safeguarding our planetary resources and their countless ecosystem services requires our willingness to risk adapting our thinking in order to gain a much wider perception of the world and its possibilities. This adaptation includes our willingness to engage in deep listening to accept one another and our contrasting points of view or frames of reference. Proactive open-mindedness supports a collaborative process of intellectual and emotional exploration of that which is and that which might be—a shared vision of a possible future.

Two other sayings are pertinent here: (1) If you do not know where you are going, any path will take you there. (2) If you stand for everything, you soon find that you stand for nothing. Would you for a moment consider flying in a commercial airplane if the pilot did not know where he or she was going, how much fuel was aboard, and roughly when you would arrive? The survival of nomadic hunter/gatherers and herders depended on their precise navigation using landmarks and the stars.

Without a vision, we relegate where we will end up to chance and external forces. When Alice met the Cheshire Cat in Lewis Carroll's 1865 book *Alice's Adventures in Wonderland*. Alice asked the Cheshire Cat,

> "Would you tell me, please, which way I ought to go from here?"
> "That depends a good deal on where you want to get to," said the Cat.
> "I don't much care where—" said Alice.
> "Then it doesn't matter which way you go," said the Cat.
> "—so long as I get somewhere," Alice added as an explanation.
> "Oh, you're sure to do that," said the Cat, "if you only walk long enough."[11]

Although a vision may begin as an abstract, intellectual idea, at some point it becomes enshrined in one's heart as a palpable force that defies explanation. It then becomes impossible to turn back, to settle for what came before, because to do so would be to allow a vital place inside us (or our communities) to whither. Few, if any, forces in human affairs are as powerful as an informed, shared vision of the heart.

4.3.2 Goals

In contrast to a vision, a goal is a longer-term statement of intent that remains until it is achieved, the need for it disappears, or the direction changes. Although a goal is a statement of direction, which may be vague and is not necessarily expected to be accomplished, it does serve to further clarify the vision statement. Such a goal might be stated as "Our goal is to make our community safer and more welcoming to cyclists to encourage exercise, reduced congestion, and help clean the air."

4.3.3 Objectives

An objective, on the other hand, is a specific statement of intended accomplishment. It is attainable, has a reference to time (generally shorter-term than a goal), is observable and measurable, and has an associated cost. Additional attributes of an objective are as follows: (1) it is an action step toward the goal; (2) it specifies a single outcome or result to be accomplished; (3) it specifies a date by which the accomplishment is to be completed; (4) it is framed in positive terms; (5) it is as specific and quantitative as possible and thus lends itself to evaluation; (6) it specifies only what, where, and when, but *avoids* mentioning why and how; and (7) it is product-oriented.

An example of an objective aimed at making our community safer and more welcoming to cyclists could include the following. By September 30, we will set up a series of consultations with our stakeholders: traffic engineers, citizens/commuters, and businesses (employers). The purpose will be to analyze residential and business-district traffic to identify the safest corridors that already have sufficient room for bike lanes and are conveniently adjacent to typical, favored routes to work, school, recreation, and so on. The stated objective is action-oriented: "We will set up a series of consultations." It has a single outcome: to locate safe, existing corridors. It specifies a time: September 30, and it is framed in positive terms. It lends itself to the evaluation of whether or not the stated intent has been achieved, and it clearly states what, where, and when. Finally, it is product- or outcome-oriented.

As one strives to achieve such an objective, one must accept and remember that an objective is fixed, but the plan to achieve it must remain flexible and changeable (because we are working with people and their schedules). A common human tendency, however, is to change or abandon altogether the objective—devalue it—if it cannot be reached in the chosen way or by the chosen time. It is much easier, it seems, to devalue an objective than it is to change an elaborate plan that has shown it will not work as originally conceived.

It is important to understand what is meant by vision, goal, and objective because collectively they tell us where we are going, the value of getting there, and the probability of success. Too often, however, we "sleeve shop." Sleeve shopping is going into a store to buy a jacket and deciding which jacket you like by the price tag on the sleeve.

The alternative to sleeve shopping is to (1) determine what you want by the perceived value and purpose of the outcome; (2) make the commitment to pay the price, whatever it is; (3) determine the price of achieving the outcome; (4) figure out how to fulfill your commitment—how to pay the price—and make a commitment to keep your commitment; and (5) act on it.

Although it is we who define our vision and goals and the objectives for achieving them, in planning it is the land that limits our options, and we must keep these limitations firmly in mind. At the same time, we must recognize that limitations can be viewed either as the solid ground on which to build new paths—or as obstacles in our preferred path. How we choose depends on how we approach life.

4.4 Barriers to Communication

In a sense, generalized personality traits are an amalgamation of the mechanisms with which one navigates life. These traits, which we each possess to a greater or lesser degree, are not cut-and-dried but rather overlapping tendencies with varying shades of gray. They can also be substantial barriers to communication and thus to the creation of a shared vision.

4.4.1 Piece Thinkers versus Systems Thinkers

There are also *piece thinkers*, the people who tend to focus on individual pieces of a system, or its perceived products, in isolation from the system itself. *Systems thinkers*, on the other hand, tend more toward a systems approach to thinking. A person oriented to seeing only the economically desirable pieces of a system (the costs and benefits) seldom accepts that removing a perceived desirable or undesirable piece can or will negatively affect the productive capacity of the system as a whole. If he or she has even thought about it, a piece thinker's typical response is "Show me. I'll believe it when I see it."

In contrast, a systems thinker sees the whole in each piece and recognizes that tinkering with the pieces might inadvertently upset the desirable functions of the interdependent system as a whole. A systems thinker is also likely to see humans as an inseparable part of the system, whereas a piece thinker normally sets humans apart from and superior to the system. A systems thinker is willing to focus on transcending the issue in whatever way is necessary to frame a vision for the good of the future. A piece thinker may need to be gradually cultivated and convinced into holistic thinking.

The piece thinker may be more reticent to (and even fearful of) change. This type of individual sees change as a condition to be avoided because they feel a greater sense of security in the known, familiar elements of the status quo. But as Helen Keller once said, "Security is mostly a superstition. It does not exist in nature. . . . Life is either a daring adventure or nothing."[12] The systems thinker is more likely to agree with Helen Keller and risk change on the strength of the abundant possibilities it offers.

A piece thinker may well be concerned primarily with land ownership and the rights of private property and wants as much free rein as possible to do as they please on their property, at times without regard for the consequences for their community, their neighbors, or future generations. Also, the greater their tendency is to place primacy on their personal needs or desires, disregarding other people and the sustainable capacity of the land. In addition, they may see their opinions as undisputable.

A systems thinker, on the other hand, is likely to be concerned about the welfare of society and all living systems, including those of the future. Systems thinkers more readily accept shades of gray in their thinking than do piece thinkers, who may see issues limited to right and wrong, black and white.

4.4.2 Abstractions versus Concrete Experiences

Some people can take ideas seemingly at "random" from any part of a thought system and integrate them; these people have mental processes that instantly change direction, arriving at the desired destination in a nonlinear, intuitive fashion. Others can think only in a "linear sequence," like the cars of a train; these people have mental processes that crawl along in a plodding fashion, exploring this avenue and that, without assurance of ever reaching a definite conclusion. If the random thinker is also at ease with abstractions but the linear-sequence thinker requires concrete examples, their attempts to communicate may well be like two ships passing in a dense fog. The result is their collaboration requires practice, patience, and commitment.

The ways in which people process and communicate information, feelings, and experiences are extremely diverse, making the creation of a shared vision a sometimes daunting but always rewarding challenge. This test brings us to the use of abstractions.

Concrete words refer to objects a person can directly experience. Abstract words, on the other hand, represent ideas that cannot be directly experienced. Abstract words are shorthand symbols used to sum up vast areas of experience or concepts with meanings that often evolve, such as "best management practices," a term used in a variety of settings and based on data and expert judgment of the day. Albeit they can be important, convenient, and useful, abstractions can lead to misunderstandings if they are not defined and described within the context of concrete examples.

The danger of carelessly using abstractions is that they may evoke an amorphous generality in the receiver's mind and not the specific item of experience the sender intended. The receiver has no way of knowing what experiences the sender intends an abstraction to include, especially when dealing with one's own core values projected into the process of creating a shared vision. For example, it is common practice to use such abstract terms as "proper method" or "shorter than" but without precise details these terms fail to convey the sender's meaning. What, exactly, is the proper method? Shorter than what?

When we use abstractions, we must be careful to link them to specific experiences through examples, analogies, illustrations, and/or actual experiences. When meeting with someone for the first time, including groups, it is often better to use, as much as possible, simple, concrete words with specific meanings. In this way, the sender gains greater control of the accuracy of images produced in the receiver's mind; the receiver gains insights and new knowledge, making language more effective in communication.

Dealing with communities in relation to their landscapes, especially cross-cultural communities or cross-cultural community interactions, it is advisable to get group members into the field, where they can physically wander through an area, such as one being considered for the open-space component of a shared vision. By using on-the-ground realities, the members can then thoughtfully discuss ideas and plans. Abstractions can thus be transformed into concrete examples, which one can experience through hearing, sight, touch, smell, sound, and/or taste to develop a common frame of reference in regard to their mission.

The importance of concrete (and common) experience can hardly be overemphasized. A major barrier to clear communication is the inability to transfer the outcomes of experience from one type of situation to another. This is especially true when people of different generations, cultural backgrounds, experiences, and knowledge about the topics come together to share ideas, including the discussion of contentious issues that arise. The potential ability to transfer results of experiences from here to there is influenced by the breadth of one's personal experiences. Every person present represents a vast array of experiences—some broad, others narrow, all different.

Experiential transfer is critical to the future, as well as the present. You may be able to recall times when a grandparent, parent, or older neighbor shared intricate knowledge about a landscape or watershed with you. How well we understand our surrounding, interconnected, interdependent ecosystems and their functions determines the quality (sturdiness) of the bridge between a community and its social-environmental landscape, as well as its present and future values and options. Hence, the ability to show how potential outcomes can be projected to a variety of possible future conditions is also crucial when dealing with the inevitable and natural disagreements that arise during a community's discussions of its sustainability.

When group members lack the necessary frames of reference, they may find the ideas presented to be abstractions, whereas those with the required experience may relate to the concepts presented more like concrete examples. Well-crafted analogies come in handy in this situation.

To make sure that an analogy will be understood, one must ask the person or people to whom one is speaking if they are familiar with the concrete example that one proposes to use in helping to extend the frame of reference to include the abstraction. If, for instance, one is talking about the value of understanding how the various components of an ecosystem interact as a basis for the system's apparent stability, one can use simple examples:

1. What happens when just one part is removed? While I (Chris) was working Nepal, a helicopter crashed and killed two people. The immediate question was what happened and why?

   A helicopter has a great variety of pieces with a wide range of sizes, which are held together by many nuts and bolts. Each nut secured to its bolt has a small sideways hole drilled through them so that a tiny piece of "safety wire" can be inserted; the ends of which are twisted together to prevent the tremendous vibration created by a running engine from loosening and working the nut off the bolt.

   Upon examination, it turned out that the helicopter crashed because a mechanic forgot to replace one tiny safety wire that kept the lateral control assembly together. A nut vibrated off its bolt, the helicopter lost its stability, and the pilot lost control. All this was caused by one missing piece that altered the entire functional dynamics of the aircraft. The engine had been "simplified" by the omission of one required piece—a small length of wire.

   Which piece was the most important part of the helicopter at that moment? The point is that each part (structural diversity) has a corresponding relationship (functional diversity) with every other part. They provide stability only by working together within the limits of their designed purpose.
2. What happens when a process is "simplified"? A newly elected mayor of a city, whose budget is overspent, guarantees to balance the budget; all that is necessary, in a simplistic sense, is to eliminate some services whose total budgets add up to the over-expenditure. A "simplistic sense" is used here because it is not quite that simple. What would happen, for example, if all police and fire services were eliminated? Would it make a difference if the price was the same and the budget could still be balanced if garbage collection were eliminated instead?

The trouble with such a simplistic view is looking only at the cost of and not the value of the function performed by the service. The diversity of the city is being simplified by removing one or two pieces or services, without paying attention to the functions performed by those services. To remove a piece of the whole may sometimes be acceptable, provided we thoroughly understand which piece is being removed and why, the piece's function and value and what effect the loss of its function will have on the stability of the system as a whole.

As the principle becomes clear the abstraction begins to take on the qualities of a concrete idea, which usually dissolves the barrier to communication and thus advances the person's ability to grasp the shared vision. Once it is certain that an analogy is understood, each member can help another person understand the abstraction. In that way, group members build valuable knowledge and experience together. This cooperation is sometimes referred to as capacity building and empowerment.

When dealing with a shared vision of social-environmental sustainability, for example, we must remember always that nature deals with trends over various scales of time. Habitat (food, cover, space, and water) is a common denominator among species (including humans). We can use this knowledge to our benefit by deeply appreciating what long-term social-environmental sustainability requires: short-term economic goals and objectives must be stated precisely, realistically, transparently, and in the *positive* with respect to the community's environmental expectations. These are the collective foundation for sound, long-term ecological goals and objectives and the resulting legacy one generation passes to another.

4.5 Reframing a Negative as a Positive

There is great power in learning to reframe negatives into positives. In so doing, the participants are creating a shared vision, not only by understanding their community from several vantage points but also understanding that much of the confusion in communication comes from trying to "flip" negatives. Trying to move away from a negative can lead to getting stuck—the group is focused on what they do *not* want. As long as people express what they do not want, the conversation can become circular and new information cannot arise. It is virtually impossible to move forward, to figure out what they *do* want.

Beginning with parents, our educational system in the United States *tries* to stress the positive. To frame our own thinking and communication in the positive as a practice—until it becomes second nature—is a worthy challenge. However, we are all exposed to the unremitting negativity daily illustrated in the news, television, movies, political discourse, and advertising, such as "don't forget" rather than "remember." What does it mean to practice framing in the positive?

Suppose your neighbor who lives along your busy street has a little boy, Jimmy. One day Jimmy's mother says to him, "Jimmy, don't go into the street." The directive words (those telling Jimmy what to do) are "don't go" (a confusing contradiction) and the last word Jimmy hears is "street." So he follows the direction of the two congruous words he hears, "go" and "street," and narrowly misses getting hit by a car.

What Jimmy's mother really meant and needed to have said was, "Jimmy, stay in the yard." Then the directive word (the one telling Jimmy what to do) would have been "stay" (a positive command, clear, and concise), and the last word Jimmy would have heard would have been "yard." He would then have followed the two congruous words he heard, "stay" and "yard," with the result of keeping safe.

This example illustrates that as hard as we try to move *away* from the negative, we cannot. Becoming aware of the negativity bias in our thinking and decision-making, we can mindfully begin to practice reframing our perspective in order to effectively move forward *toward* the positive. Negativity bias is defined as "an automatic tendency to pay significantly more attention to unpleasant than pleasant information." In other words, negative events affect our behavior more than positive events.[13]

In my (Holly's) experience as the leader of two, small research centers and supervisor of small teams, it is ineffective to begin a sentence with "don't"—a lesson I have used ever since across many settings including teaching, family, and other relationships. Similarly, I was trained to not begin a sentence with "you," as in "You always [fill in the blank]" (e.g., "You always say that," "You always leave your shoes in the middle of the hall," "You always fidget when we are having a conversation"). Among other barriers, using "don't" and beginning sentences with "you" can sabotage communication by inadvertently putting the listener on the defensive, in which case they stop listening immediately. Alternatively, asking a helpful/neutral question is a far more respectful and effective means of communication, whether conversing with children, adults, students, or employees. "Have you thought of approaching the task in another way?" This affords respect to the person(s) whom you are addressing and simultaneously avoids falling into the communications trap represented by the fact that adults do not like to be told what to do (unless their immediate safety is at risk). Asking a helpful/neutral question also elicits the listeners' own ideas or suggestions toward solutions so they are actively involved and feel valued on a team and take ownership of solutions.

Negativity begets stuckness or circular thinking. Being stuck because you are attached to what you *do not* want holds you captive from discovering what you *do* want and then establishing the steps toward achieving it. This brings us to the notion of constraints.

4.6 A Vision Is Tied to Negotiability Constraints

The vision of some future desired condition, by its very nature, elicits the singular social constraint (the fixed point around which everything else turns, like the hub of a wheel) that must be met if the terms of the vision are to be fulfilled. A constraint in this sense means being restricted to a given course of action or inaction, which connotes something that restricts, limits, or regulates personal human behavior.

A vision does not in and of itself create a single constraint where there was none before. Any decision to choose an option for moving forward presents various constraints. Everything in the world is already constrained by its relationship to everything else, meaning that nothing is ever entirely free. A vision determines the degree to which a particular socially chosen constraint is negotiable. What course of action are we committing ourselves to in order to achieve our vision? The power of the vision rests with its authors—the people who created it and those who are inspired by it, not those whose sole job is to administer the details.

What does "negotiable" mean here? It means to bargain for a different outcome, to cut the best, acceptable deal possible. For example, most changes in climate are determined by nature and are nonnegotiable. Can we negotiate with nature to give us sunnier, drier winters without flooding when we deem the winters too dark and wet? Can we negotiate for more rain in an area of drought? Well, we could try, but it would be to no avail. Nature does not negotiate; therefore, some of the conditions nature hands us are nonnegotiable; we cannot "cut a better deal," one more to our liking.

Our challenge, therefore, is to learn what is negotiable and what is not. Beyond this, we must learn to accept with grace that which is non-negotiable and learn to account for and accept responsibility for the price of that which is negotiable, because negotiability is not free. We are simply trading one set of behavioral freedoms for another in that we impose a particular constraint on ourselves through a vision in order to alleviate some other potential constraint in the future—the desired outcome of our vision.

Is there anything that we humans can negotiate amongst ourselves wherein nature does not have the final word? To a small extent, yes, but with the caveat that the outcome of our behavior must not breach one of nature's irrevocable, irreversible biophysical principles.

As a society, we can negotiate the rules and regulations of our self-governance, our economic theories, conditions, and chosen practices. And it is exactly because we are able to negotiate these societal norms that their degrees of negotiability are also open to those who would bargain for a better deal. Hence, the question is this: once we create a shared vision, how does it affect the negotiability of our self-created, self-imposed theories, rules, and regulations?

A shared vision determines, by its defined outcome, the degree of negotiability that can be afforded to everyone who is subject to our relevant self-regulation, our economic theories and conditions, and our chosen practices. In other words, a vision determines the negotiability of any particular social (behavioral) constraint, with which we have to deal in everyday life, be it how one interprets the rights of private property, how one conducts oneself in a religious gathering, or where one chooses to build one's house—in the floodplain of a river or on high ground.

In summarizing the negotiability issue, we are asking the following: Will we, as a community, be able to negotiate the changes we are committing to through our community vision? More open space, more bike lanes, more access to carpooling and public transport, more green space, more community gardens, and throughout the issue of broader accessibility: we will *all be able to partake* of the vision and its components regardless of our age, income, health and physical ability, and so forth. *The final community vision is equitable and fair and benefits everyone, not just some.*

4.7 Testing the Effectiveness of Decisions through Monitoring

Although the word "monitor" is variously construed, its meaning here is to scrutinize or check systematically with a view to collecting specific kinds of data that indicate whether or not one is moving in the direction one wishes. This is also referred to as adaptive management: we implement a plan, and we evaluate it and report our findings as we go along.[14] Monitor has the same origin as "monition," which means a warning or caution, and is derived from the Latin *monitio*, "a reminder."

With respect to the biophysical sustainability of nature's landscape, the foundation of every community, monitoring means keeping watch over human activities throughout a long period of time in order to collect data, such as how, when, and why individuals maintain the desired course or stray from that course. Monitoring informs our planning and helps us understand how our community is negotiating the vision. The results of monitoring help us conserve the opportunities embodied within the system for ourselves and future generations. Because planned observation is dependent on questions, it requires the ability to design precisely targeted, relevant questions.

4.7.1 The Questions We Ask

Learning how to frame effective questions is paramount not only for crafting a collective vision but also for the process of monitoring what is necessary to achieve that vision over time. A thoughtfully composed question is a powerful tool when introduced strategically because questions open the door of possibility. These are sometimes referred to as probing questions, and rather than being complex, they may be stark and straightforward and yet be incredibly thought-provoking and effective. For example, it was not possible to go to the moon until someone asked, "Is it possible to go to the moon?" At that moment, going to the moon became possible. To be effective, each question must (1) have a specific purpose, (2) contain a single idea, (3) be clear in meaning, (4) stimulate thought, (5) require a definite answer to bring closure to the human relationship induced by the question, and (6) explicitly relate to previous information.

In a discussion about going to the moon, one might therefore ask, "Do you know what the moon is?" The specific purpose is to find out if one knows what the moon is. Knowledge of the moon is the single idea contained in the question. The meaning of the question is clear: do you or do you not know what the moon is? The question stimulates thought about what the moon and may spark an idea of how one relates to it; if not, that can be addressed in a second question. The question asked is both related to the previous information and requires a definite answer.

A question that focuses on "right" versus "wrong" is thus a hopeless exercise because it calls for human moral judgment, and that is not a valid question to ask: who is the arbiter of morals for us all? A more effective question might be whether a proposed action is "beneficial" or "detrimental" in terms of a community's collective vision.

For example, a good short-term economic decision may simultaneously be a bad long-term ecological decision and thus a bad long-term economic decision. To find out, however, one must ask this: although this is a good short-term economic decision, is it also a good long-term ecological decision and hence a good long-term economic decision? An answer to anything is possible only when the question has been asked. Once again, we are seeking to advance an inclusive, informed vision ultimately based on a decision process that simultaneously helps us achieve the three pillars of sustainability: social, ecological, and economic sustainability.

In essence, questions lead to an array of options from which one can choose. Conversely, without a question, one is blind to the options. Monitoring helps us learn about the options. In turn, knowing what, when, and how to monitor requires understanding what questions to ask.

4.7.2 Monitoring in Five Steps

Within the context of constant change and future uncertainties, the best monitoring and adjustments (target corrections) provide several preplanned, logical actions, such as: "If A happens, I will do B; if C happens, I will do D;" and so on. Otherwise, we monitor only outcomes and merely hope corrective actions will be found in time if we observe the outcomes to be undesirable.

An example might be the assumption that placing fish ladders in dams will sustain the migration of salmon and perpetuate their survival. Over time, however, we learn that the reservoirs created by the dams (in addition to a host of other human activities) also affect the survival and migration of the salmon. Therefore, monitoring a single variable, even a seemingly reasonable one, such as counting fish, may do nothing to clarify the issue or save the fish, which was the purpose of the fish ladders.[15] We must therefore accept and remember that things are inevitably more complex and interrelate than we are able to foresee, which is the very reason we need to monitor in the first place. In order to understand the complexity, visualization is necessary.

Visualization, beyond taking our groups out to the sites we are discussing, is a major tool available to us for getting at the complexities of our project and our ultimate options for implementing the vision.

From the beginning of our conversations and at any point in the visioning process the group and its facilitator(s) may find it helpful to examine photographs and maps and for them to create drawings and diagrams, such as concept maps. This can start with a stack of paper at the tables, a large pad of paper hanging on a frame, or several whiteboards around the room for the people to list their wish lists, questions, or desired outcomes. Groups should regularly try visualization exercises when brainstorming. It helps members understand each other's ideas. The visioning process benchmarks from week to week and the where-are-we status of the process are led by the facilitator. Visualization can also give rise to critical questions that help us proactively brainstorm potential flaws or pitfalls in our plan, unforeseen circumstances, or unintended consequences to be avoided to the greatest extent possible—affording troubleshooting that saves heartache, headaches, time, and money.

Group members new to visualization and diagraming will soon discover that it also helps with daily decisions (jobs, apartments, houses, cars, etc.) In addition, it is fun and provides a variety of communication modes to the meetings: it gets members up on their feet, walking around the room and having smaller-scale discussions and breakthroughs. Visualization can be employed at every step in the process, including monitoring.

Monitoring has five steps: (1) crafting a vision, goals, and objectives; (2) preliminary monitoring or inventory; (3) monitoring implementation; (4) monitoring effectiveness; and (5) monitoring to validate the outcome(s).

4.7.2.1 Step 1: Crafting a Vision Statement, Goals, and Objectives

Crafting a carefully worded vision and attendant goals and objectives, which state clearly and concisely your desired future condition and how you propose to get there within some scale of time, is the necessary first step in monitoring. This beginning clarifies what you want, where you want to go, and what you think the journey will be like. The vision, along with its goals and their objectives, form the context of the journey against which you measure (monitor) all decisions, actions, and consequences to see if, in fact, your journey is even possible as you imagined it and what the consequences of the journey might be.

Once you have completed your statements of vision, goals, and objectives, you not only will be able to but also must answer the following questions concisely: (1) What do I want? (2) Why do I want it? (3) Where do I want it? (4) When do I want it? (5) From whom do I want it? (6) How much (or how many) do I want? (7) For how long do I want it (or them)? If a component is missing, you may achieve your desire by default but not by design.

Only when you can answer all of these questions concisely do you know where you want to go, the value of going there, and if you are able to calculate the probability of arrival. Next, you must determine the cost in both monetary and nonmonetary terms, make the commitment to bear it, and then obligate yourself to keep your commitment.

4.7.2.2 Step 2: Preliminary Inventory

To take inventory is to carefully observe and understand the circumstances that you begin with, which means accounting for what is available in the present. This establishes our baseline for decisions. If the landscape is one for which we have data going back, we can identify trends also. It would be important to acknowledge these trends before placing the overlay of our purposes and desires on top of it.

Taking inventory requires these questions: (1) What exists now before anything is purposefully altered? (2) What condition is it in? (3) What is the prognosis for the future? Even though monitoring may require multiple questions, the outcome is still a single realization.

If, for example, you go to your doctor for an annual checkup, the doctor would have to take a series of measurements, such as your blood pressure and blood tests, and would have to know what a healthy person is (including, if possible, you as a healthy person), as a benchmark against which to evaluate your current condition. If you are indeed healthy, then all is well; if not, your doctor would presumably prescribe tests to pinpoint what is wrong and ultimately prescribe lifestyle changes, such as diet, exercise, and medicine to improve your condition and make a prognosis for your future.

If, on the other hand, you go to your doctor but only allow him or her to take your blood pressure—without checking other components, such as the level of your cholesterol—your doctor cannot deal with your health as a systemic whole and thus loses the ability to see the various components as parts of an interactive, interconnected, interdependent system. In this, your body is similar in principle and function to your family, the community in which your family resides, the landscape in which your community rests, and the landscape within the bioregion.

4.7.2.3 Step 3: Assessing Implementation

Assessing the implementation of projects on the ground asks the following question: Did we precisely do each step we said we were going to do?

4.7.2.4 Step 4: Monitoring Effectiveness

Monitoring to assess effectiveness is focused on the implementation of your objectives, not the goals or vision. Recall from the earlier discussion that a vision and its attendant goals describe the desired future condition for which you are aiming. They are qualitative and thus not designed to be quantified. An objective, on the other hand, is quantitative and so is specifically designed to be quantifiable.

To assess the effectiveness of an objective requires asking the following: (1) Is the objective specific enough? (2) Are the results clearly quantifiable and within specified scales of time? Monitoring the effectiveness of your project with the aid of indicators provides information (feedback) with which to assess whether you are (1) headed toward the attainment of your desired future condition (the condition of your collective vision), (2) maintaining your current condition, or (3) moving away from your desired future condition. Monitoring always includes asking when, how often, and for how long. For example, monitoring stream temperatures and pH does not tell us much without our knowing when, how often, and for how long the stream was monitored so we can compare results over time within all sorts of conditions or seasons. How long it will take to achieve the objectives must be factored in; the estimate of how long must be realistic or otherwise conservative. Group members (and stakeholders) informed of a realistic timeline will follow the progress with great interest, no matter how long it takes. Conversely, group members and stakeholders may lose interest in any future projects or become vocally opposed if the project takes far longer than the estimate indicated. Will it take three years or thirty to revitalize the banks (the riparian zone) of a particular reach of a river?

Assessing effectiveness is the systematic monitoring of indicators that are relevant to achieving your vision. A good indicator helps a community recognize potential problems and provides insight into possible solutions. What a community chooses to measure, how it chooses to measure it, and how it chooses to interpret the outcome will have a tremendous effect on the quality of life in the long term.

Indicators close the circle of action by both allowing and requiring that you return to your beginning premise and ask (reflect on) whether, through your actions, you are better off now than when you started: If so, how? If not, why not? If not, can the situation be remedied? If so, how? If not, why not? And so on.

Here, caution is necessary. Traditional, unidimensional indicators, which measure the robustness of one condition (say the economy), ignore the complex relationships between economies, communities, environments, and neighboring communities and within the bioregion. When each component is evaluated as a separate issue and thus evaluated in isolation, measurements tend to become skewed and lead to ineffective policies, which in turn can lead to a deteriorating quality of life. Indicators must therefore be multidimensional and must represent and measure the quality of relationships among the components of the system being monitored if a community is to have any kind of accurate assessment of its sustainable well-being.

Only with relevant indicators and a stepwise, systematic way of tracking them is it possible to make a prognosis for the future based on your vision, goals, and objectives, which states a desired future condition within some scale of time and a plan to achieve it. Only with relevant indicators and a systematic way of tracking them is it possible to make the necessary target corrections to achieve your vision, because only now can you know what corrections to make.

4.7.2.5 Step 5: Monitoring to Validate the Outcome

Monitoring may seem to be the weakest point in how a community deals with its shared vision for a sustainable future. Yet the crux of making a shared vision work lies in monitoring, which requires many contributors and much time, thought, input, and commitment to follow through. A shared vision for a sustainable future must be considered a working hypothesis because it is applied to ongoing and proposed human activities within a dynamic ecosystem whose multiple interactions and self-reinforcing feedback loops are largely unknown and whose outcomes are, therefore, uncertain.

Because outcomes of a shared vision are uncertain, human activities encompassed by a vision can be thought of as tentative probing into various aspects of nature and are best taken one step at a time and tested at each step. Through such testing, one hopes to detect potentially adverse and unpredicted effects as early as possible, to troubleshoot through adjustments to the community's best ability to prevent serious, widespread, or irreversible damage.

Although adaptive management can be likened to a weak flashlight beam guiding us through the inky dark night (because it is not perfect, and it is not perfect because ecosystems are incredibly dynamic), the same can be said of a shared vision projected into a hoped-for future in which the weakest part of the beam is the information feedback system.[16] Clearly, it is one thing to promise a complicated, expensive, and continual activity, such as repetitious monitoring to create a continual flow of information (particularly for the effectiveness of actions and causative trends) between the people of a community and their immediate environment, but it is quite another to actually implement it as promised.

4.7.3 Putting Time in Perspective

Beyond the time and money committed to doing the visioning process correctly, communities spend the preponderance of their time, energy, and money in planning, which is carried out in the sociopolitical-economic arena, where many of the basic dimensions of monitoring are formulated. It is at this point that the scientific basis for planned actions is reviewed, and the unknown effects of the planned actions on the environment elucidate the often-controversial nature of a community's proposed activities—despite the fact that the character, importance, and probability of unwanted ecological surprises in the future are seldom explored or made explicit. Here, again as before, thoughtful and strategic questions (how? what if?) help shape successful outcomes and avoid foreseeable problems.

Be that as it may, to garner the support of skeptical or unreceptive members of a community, planners may tell the members that an activity will either be modified or stopped (if scientific information warrants it) either at the earliest sign of an adverse effect or when it becomes clear that the specified objectives will not be met. This is where well-informed, proactive troubleshooting prevents costly mistakes and backtracking when it is firmly part of the collaborative process at the beginning of the project vision/plan. Some specified objectives may be met sooner than others. This is the nature of working with an ecosystem, as the *iterative* process unfolds it is teaching us valuable lessons. Moreover, no two parcels of land, rivers, aquifers, or projects are alike; each is utterly unique and there is only a very rough pattern book of data for us to follow. In some regions and systems, data to help inform our plan may be scant or nonexistent. This lack of information requires taking our time, direct observation, the use of the closest analogs we have (such as from an adjacent region), and monitoring.

One of the major faults associated with the implementation of a shared vision is that it is based on short-term or high-frequency responses from the environment, whereas the effects of human behavior ripple into the distant future, often beyond the ability of short-term corrective actions to be of value. In the mere instant of short-term monitoring, we find no viable answer to Garrett Hardin's question: "And then what?"[17]

Surprises (including unwelcome ones) are inevitable, and their seeds may or may not be entrained in the stream of data from monitoring and may or may not be discovered by those who observe the data because most monitoring programs are scaled to the immediate future, not the slower, longer-term responses of the environment. Planners who want quick feedback may feel tempted to promise corrective actions only in the near term or slightly distant future. Too often planners become mired in a cycle in which they react from crisis to crisis (from ecological surprise to ecological surprise) in part because they do not see the correlation between information that occurs frequently and events at the next magnitude of scale that occur less frequently than the information would indicate.

This is not to say that monitoring short-term events is unimportant but rather that only considering such events in ecosystems with long-term feedback loops and/or high degrees of environmental variability will likely prove to be a disastrously myopic choice. Increasingly, long-range studies are one tool that overcomes this issue (for example, long-term ecological research, or LTER, may study a site over decades[18]). However, the very prospect of monitoring the long-term ecological effects of human activities may seem improbable, if it were to successfully get started at all, because humans have been known to have a short attention span leading to broken, informational feedback loops.

4.7.4 The Danger of Interrupting Information Feedback

The most important feedback in response to environmental information occurs when planners decide whether they should change the direction or intensity of a particular activity. Because this decision requires the longest environmental feedback loop, it is the one most likely to be broken without a firm, serious commitment by the entity in charge (nation, province, state, city, county, or agency), which instructs the planners and the community citizens themselves. This requires a significant, lengthy commitment by the staff (and citizen science) with adequate funding to complete the task of long-term monitoring. Things to be aware of and pitfalls to avoid include the following:

1. Planners, members of the city council, the mayor, and/or county commissioners lose interest and hope the issue will fade or the crisis will wane.
2. Monitoring (especially long-term) is too expensive and either loses or never had adequate funding. Diminishing both the level of funding and the number of trained personnel is perhaps the most frequent cause of failure.
3. The lag time between implementing an activity and the environment's response (known as hysteresis) is so long that a planner gets no data, stale data, messy data, or data that are inappropriate for timely decisions or actions.
4. The issue(s) driving the monitoring program is no longer relevant to the future citizens of the community.
5. The system of monitoring apparatus (such as photo sites, corner stakes, and other markers identifying permanent sampling plots) is not maintained, even though interest in the issue may rekindle. The most serious loss to effective monitoring is the inability to return to the *exact* field locations for repeat measurements. Such loss can be overcome with Global Positioning System (GPS) and a geographic information system (GIS) that people can even access on their phones

or with professional GPS equipment, such as a Garmin. These applications can then be used over time to create illustrative maps. In the absence of habitat-sensitive or migrating species, even drones equipped with GPS and cameras can greatly aid in the gathering of data, forming the basis of outdoor education and outdoor citizen science, both with the presence of expert assistance.

6. There are too many issues to monitor every single one, so the community and planners must establish priorities linked directly to the project's objectives, or the monitoring program may fade into history before fruition.
7. Monitoring may also fail if an issue emerges that is too hot to handle, so *no action* is deemed better than potentially placing oneself in a politically vulnerable or losing situation. People can disagree about an issue, and if the numbers are fragmentary or incomplete, both can be "right." In highly controversial situations, the planner, city councilor, mayor, or county commissioner are keenly aware they may be replaced. Furthermore, it is almost always a lose-lose situation when that which is to be evaluated has already been irreparably degraded prior to the onset of monitoring.
8. There may be a reluctance on the part of present planners, city councilors, mayors, and county commissioners to commit their future replacements to present obligations. But are the present questions really crimping the generations of the future who would benefit by continuing to collect the data to build upon the data we have bequeathed them, as an unconditional gift of potential knowledge? Will they not have the opportunity to find better ways of addressing the issue(s)? Will today's issues even be relevant in that distant time?

Any of these eight circumstances weaken the level of commitment and severs the flow of information, which thereafter cannot be repaired. Bear in mind that these maladies discussed are at least partly related to who funds the monitoring; who develops, conducts, analyzes, and interprets the data; and who makes the decisions about subsequent adjustments in activities. A diverse set of collaborative partners is an investment in success. The partners may include agencies, scientists of all relevant disciplines, educators/schools/universities, business and nonprofit leaders, foundations, and grantors from across the aforementioned spectrum.

It is the questions we ask that frame our perceptions (and vice versa). In the end, it is through the questions we ask that we derive and finely craft our vision, goals, and objectives and direct our monitoring and adaptive actions. Social-environmental sustainability depends first and foremost on the relevance and precision of the questions, for they become our compass and map into the future.

As community leaders, let us facilitate group processes that transform a collection of individuals into a healthy, sustainable community within the context of a productive, sustainable environment in the present—for the present and the future. Let this be our personal legacy to the following generations. After all, "the care of the earth is our most ancient and most worthy and . . . our most pleasing responsibility," says Wendell Berry. "To cherish what remains of it, and to foster its renewal, is our only legitimate hope."[19] But this hope depends on, hinges on, wise leadership, commitment, and involvement on the people's part and the type of steadfast determination to see the vision through and shape it, as it unfolds.

4.8 References

1. The discussions of values and aspects of a vision follow, in part: Laurence G. Boldt. 1993. *Zen and the Art of Making a Living.* Penguin Books, New York, NY., 608 pp.
2. Chris Maser, Russ Beaton, and Kevin Smith. 1998. *Setting the Stage for Sustainability: A Citizen's Handbook.* Lewis Publishers, New York, NY., 275 pp.
3. Edmund Burke. 1899. Letter to a Member of the National Assembly 1791. *The Works of the Right Honorable Edmund Burke*, 4:51–52.
4. Will Steffen, Kartherine Richardson, Johan Rockström, and others. 2015. Planetary Boundaries: Guiding Human Development on a Changing Planet. *Science.* DOI: 10.1126/science.1259855. www.mentalfloss.com/article/19206/clock-long-now 347(6223).
5. Albert Einstein ibid. Chapter 1 Note 6.
6. Suzy Kassem. 1975. Quote may be from her book, *Rise Up and Salute the Sun: The Writings of Suzy Kassem.* Awakened Press, ISBN 0615388868, May 15, 2011.
7. Plato, in the Republic Stated That "Our Need Will be the Real Creator." Some Believe This is the Origin of the Proverb "Necessity is the Mother of Invention." *See Design@Open.* www.open.ac.uk/blogs/design/necessity-is-the-mother-of-invention/.
8. Byron Katie and Stephen Mitchell. 2003. *Loving What Is: Four Questions That Can Change Your Life.* Harmony Books, New York, NY., 352 pp.
9. Albert Einstein. www.brainyquote.com/quotes/albert_einstein_384440 (accessed January 28, 2022).
10. Yeats William Butler. 1916. *Responsibilities: And Other Poems.* Macmillan, Ireland.
11. Lewis Carroll. 1933. *Alice's Adventures in Wonderland.* Doran, & Co., Doubleday, New York, NY., 162 pp.
12. Helen Keller. 1957. *The Open Door.* Doubleday, Garden City, NY., 140 pp.
13. Zhuo Jing-Schmidt. 2007. Negativity Bias in Language: A Cognitive-Affective Model of Emotive Intensifiers. *Cognitive Linguistics*, 18(3):417–443.
14. B. K. Williams, R. C. Szaro, and C. D. Shapiro. 2009. *Adaptive Management: The U.S. Department of the Interior Technical Guide.* Adaptive Management Working Group, U.S. Department of the Interior, Washington, DC., pp. 72:4.

15. Chris Maser. 2004. *The Perpetual Consequences of Fear and Violence: Rethinking the Future*. Maisonneuve Press, Washington, DC., 373 pp.
16. L. Rist, A. Felton, L. Samuelsson, C. Sandström, and O. Rosvall. 2013. A New Paradigm for Adaptive Management. *Ecology and Society*, 18(4):63. http://dx.doi.org/10.5751/ES-06183-180463.
17. Garrett Hardin. 1986. *Filters Against Folly: How to Survive Despite Economists, Ecologists, and The Merely Eloquent*. Penguin Books, New York, NY., 247 pp.
18. National Science Foundation (NSF). *Long Term Ecological Research and Examples of Discoveries*. www.nsf.gov/news/newsmedia/ENV-discoveries/LTER-discovery-series.jsp (accessed January 13, 2023).
19. Wendell Berry. 1993. *Sex, Economy, Freedom, and Community*. Pantheon Books, New York, NY., 208 pp.

4.9 Discussion Questions

1. Compare and contrast the concepts of a vision, goals, and objectives. What are the differences between a person's perspective and their perception?
2. What are the differences between linear cause-and-effect thinking and systems thinking? How might the differences influence the process with respect to formulating a community's vision of social-environmental sustainability?
3. Can the two approaches of linear cause-and-effect thinking and systems thinking be used in complementary ways in constructing a community's vision of social-environmental sustainability? How?
4. Why is it important in creating a community's vision of sustainability to translate abstractions into concrete experiences?
5. In dealing with a vision's goals and objectives, what is the difference in focusing on efficiency versus effectiveness, and how does the choice affect the outcome? Do we need to consider both? If so, why?

5

The Essence of Leadership

5.1 Introduction

An effective leader who leads intentionally must have the same characteristics as one who leads intuitively. What are these characteristics?

Psychotherapy and leadership both deal with the care of people's emotions. The personal qualities that make a fine leader are similar to, if not the same as, the qualities that make a fine psychotherapist; both must first become as healed or as whole as possible if they are to be a model worth emulating, even temporarily.

The major difference between them is the end toward which they apply principles and insights. A psychotherapist helps people to reassess and reassemble their lives on an individual basis, whereas a leader helps people to envision or reassess their future and heal their community through gradually transforming goals and relationships.

A leader is one who values and serves people and helps them transcend their fears so they might be empowered to grow beyond where they previously were. This is the essence, the first rule, of true leadership. One might translate this to leadership as follows: Rule 1: leadership is service to others based on strength of character. Rule 2: first, learn rule 1. This concept is critical because leadership has to do with authority, which is the ability to exert control, command, enforce laws, exact obedience, determine, or judge. Here, it is necessary to distinguish between leadership and managership.

In short, a leader consciously knows the right thing to do, while a manager pays minute attention to detail, to the letter of the law, and to doing the thing "right," even if it is not the "right" thing to do. A manager relies on the application of external (often tried and true, conventional) techniques to solve problems and is concerned that all the procedural pieces are both in place and properly accounted for, hence the epithet "bean counter."

Good managers are sometimes placed at a disadvantage when put in positions of leadership if all they are able to do is rise to their level of incompetence.[1] Similarly, a leader placed in the position of managership may be equally inept because the two positions require vastly different skill sets.

DOI: 10.1201/9781003353744-5

Thus, an effective leader can guide the process of sustainable community development, while an effective manager keeps it running smoothly.

In both cases, the authority of a person begins as an internal propensity that comes from the person's heartfelt belief in their higher consciousness, which can act as a reliable guide in life. As stated in the Old Testament (Proverbs), "As a man thinketh in his heart, so is he."[2] In the end, it is the collective heart of the people that counts; without people, there is no need for leaders. Chinese philosopher, poet, and founder of Taoism, Lao-tzu, thought of it this way: "A leader is best when people barely know he exists, when his work is done, his aim fulfilled, they will say: we did it ourselves."[3]

5.2 True Leadership

A person who only has the authority of position may have a socially accepted seat of power over other people, but such power can exist only if people agree to submit to the authority. Someone who holds a position of authority yet does not lead from emotional intelligence and their conscience can only manage through self-serving coercion and fear[4]—but cannot lead. As American mathematician Norbert Wiener is sometimes quoted as observing, "A conscience which has been bought once will be bought twice,"[5] to which historian John Edward Acton (1834–1902) might have added, "Power tends to corrupt, and absolute power corrupts absolutely."[6]

Each person's truth is personal; no one knows what is "right" or "true" for anyone else. Leadership is not meant to be a form of indoctrination, a method to manipulate people so they will respond in a predetermined or "right" way. To discuss the outer processes of leadership without addressing its inner essence is a hollow exercise that achieves little in the end.

The inner essence of leadership begins with the honesty of the person, to which others respond. In responding, the group accepts and validates their leader's authority. Moreover, the best leaders are people with a fine balance between the masculine and feminine aspects of their personality. Leadership means investing in people and teams first and putting relationships ahead of immediate achievement. Service is an attitude and an ethic, not a function. Leaders support their group and the larger enterprise from moral conviction, usually expressed as enthusiasm, which causes people to want to follow with action.

True leadership is a balancing act between the emotional and intellectual components, giving equal weight to each. By blending the heart and the head, leadership is raised to a higher level of ability, clarity, authenticity, and consciousness. Good leadership requires a great deal of humility.

A leader cannot keep values and beliefs out of the relationship with those whom they would lead; to do so renders their leadership distant and out of touch. Although a leader must be willing to openly discuss the issue of values, it needs to be understood that the leader's personal values will at times be revealed, which must be done with honesty, grace, and respect for the different values among the team members.

5.3 Personal Values

Unfortunately, some well-intentioned leaders are overzealous in "straightening people out," which implies that, by virtue of their "greater wisdom," they alone will dictate the answers for the group, but leadership is a process and a long-term conversation; a leader must not simply accept whatever the group members say by remaining passive and silent. When a leader senses that one option toward a goal is unwise or destructive, it becomes their duty to invite the people to examine the tradeoffs, pitfalls, and consequences of their actions, present and future.

A central issue in leadership is how the leader's values enter into their relationship with their group and with the greater public. Contemporary, effective, and thoughtful leadership is about listening, collaborating, summarizing, and presenting options to arrive at mutually acceptable goals and sustainable decisions, then following through.

An emotionally intelligent leader encourages their team to honestly take account of their professed values versus mindlessly following outmoded conventional thinking or social conditioning without conscious acknowledgement of the difference. Leaders encourage teams to be imaginative and creative and challenge the status quo by thinking outside the box in problem-solving; they help instill knowledge, capability, confidence, and capacity in teams and the public decision-making process.

At this juncture, it is important to emphasize that, because a leader's personal values do influence their relationship with the group, their honesty and self-awareness are crucial.

In order to test their own thinking, citizens deserve to understand a leader's values and the fact that neither the leader nor anyone else has all the answers. Leadership must be a process whereby the people are challenged to honestly evaluate their own values; learn about the issues, choices, and outcomes; and then decide the direction of the project for themselves.

Leaders must recognize their personal attributes, negative and positive, and seek to manage any internal conflicts or unfinished business, as well as their own sense of vulnerability or tendencies toward defensiveness. They must also recognize how these traits can interfere with their constituents being free to fully explore new ways of conceptualizing and planning.

This self-awareness and this sensitivity to others are critically important. Without these traits, a leader may inadvertently hold the group back and obstruct the group's growth and capacity. Leadership must focus on what is best for the people and the project and never be self-serving.

All people have blind spots or distortions of reality. It is therefore a leader's responsibility to continually expand self-awareness to proactively address bias, prejudice, and defensiveness. It is important to become aware of attitudes or tendencies stemming from potentially unfinished emotions in order to lead with integrity and openness toward the group's betterment rather than stagnation.[7]

5.4 The Required Characteristics of True Leadership

Educator and author John Holt (1923–1985) wrote, "The true test of character is not how much we know how to do, but how we behave when we don't know what to do."[8] Too often, people who lack an inner compass, dedication to group process, and the willingness to take risks characteristic of true leaders find themselves in positions of power. Conflating the trappings of power with leadership can lead to abuses of the limited authority of the position and a frustration of progress for the group. Such people are obsessed with the perceived status of superficial power and are loath to relinquish it, even momentarily. People cannot become effective leaders out of negativity, self-service, or reactivity. True leadership requires a deep reservoir of trust, a clear sense of interdependence and the value of community, a clear commitment to principle, and the courage to stick with the process and the patience it requires despite personal costs.

5.4.1 Authenticity

A true leader is concerned primarily with facilitating someone else's ability to reach their potential as a person by helping them develop their talents and skills and value their experiences. Authentic leadership comes from the heart and deals intimately with human values, human dignity, and ultimately the prospect of human development, equality, and well-being.

Authenticity is the condition or quality of being trustworthy and genuine. Beyond the dictionary definition, authenticity is the harmony between what a person thinks, says, and does and what they really feel—the motive in the deepest recesses of their heart. The adage "Deeds speak louder than words" is true as far as it goes, but what is left unsaid is that "Motives speak louder than deeds." A person is authentic only when their motives, words, and deeds are in harmony with their attitude.

Essayist Ralph Waldo Emerson (1803–1882) wrote, "Don't say things. What you are stands over you the while, and thunders so that I cannot hear what you say to the contrary."[9] A person's attitude is the visible part of their behavior, but their motive is often hidden from view. When a person's visible behavior is out of harmony with their motive, that attitude points to a hidden agenda. Therefore, an authentic person is one who is willing to risk shedding stereotypical roles in order to act genuinely—being a real person actively and attentively present in a relationship.

If a leader models realness by engaging in appropriate self-disclosure, they can anticipate that such authenticity will inspire a greater following based on trust and emulation than would otherwise be possible. In the end, the degree of aliveness, commitment, and psychological soundness of a leader (who has worked through their own issues) is the crucial variable that determines the outcome of leadership. To accomplish this, however, one must have honor.

5.4.2 An Honorable Person

Personal honor is a trait one cultivates over a lifetime and a gift one gives to oneself; it is a code of ethics and personal integrity by which one lives and interacts with others that is maintained without legal or other obligation. A leader must understand that personal integrity includes each commitment made to oneself and that such commitments are every bit as important to keep as those made to constituents. It is that inner essence, the spiritually backed standard, that defines who one is—one's compass.

A leader must have an inner sense of honor, such that one's word is one's bond. There was a time when a contract between two people was sealed with a handshake and each party's word was at least as sound as a legal document—or even more so.

Israeli prime minister Yitzhak Rabin was a shining example of a man with personal honor because he was through and through a warrior, but one who had the courage and wisdom to see that peace was the only road to the security of his country—and he chose to take that road no matter how difficult. While honor was formerly referred to in regard to gentlemen, it is an attribute shared equally between men and women, which brings us to the balance between a leader's masculine and feminine aspects.

5.4.3 Balancing One's Outlook

Although a leader can help guide us from here to there in the outer, physical-spatial landscape without actually having been there before, a leader can only *lead* us to a given place of consciousness (the inner landscape) if they, in fact, have been there. For example, society may still emphasize linear, left-brain thinking, once held as a masculine trait, over right-brain or intuitive thinking, once held as a feminine trait. However, these false dichotomies are finally breaking down.

While tapping into the so-called right side of the brain is at times more powerful than using the so-called left side, it is difficult to appeal to the right side because we may be less accustomed to or comfortable with accessing our intuition. Drawing on intuition is less prevalent in Western society, where relationship and creativity have been too often stifled by patriarchal conformity. Therefore, to appeal to the intuitive, conceptual side, a leader must either have a strong innate balance or must consciously develop it before being able to model it for others. Naturally, we need to be highly developed both conceptually and analytically—in order to engage in holistic thinking.

It is a necessity for a leader to be well balanced between conceptual and analytical aspects of understanding and problem-solving. It is critical to be able to connect emotionally with others in order to become an effective leader. While one cannot change others' minds, we can create a space where others can bring their convictions and values to the surface and share them. Before this is possible, however, the leader must actively cultivate an emotional connection, and that lies within the realm of the conceptual and intuitive.

A common human trait is to make decisions based on emotion and justify them with "fact." This same principle applies to concepts and ideas; we buy the emotion before we accept the facts, which resonates with the phrase by American author Charlotte Perkins Gilman (1860–1935) that "a concept is stronger than a fact."[10]

Therefore, leaders who touch people emotionally make a lasting connection. Such leaders not only create experiences for their constituencies or cause a long-dormant memory to be vividly recalled but also make a memorable connection that has the power to motivate.[11]

Bear in mind, however, that there is a wide range of positive and negative emotions. We are not talking here about the manipulation of others through emotions, such as fear or loathing.

Before groups will accept a new idea, the leader must be a likable person to whom they can relate. Do we feel good about him or her? Is he or she authentic and trustworthy? This is critical, because genuine passion is not only rare but also the way to the emotions and the soul, fertile ground for developing deep relationships. Trustworthy leaders are often described as dynamic, charismatic, riveting, or engaging because the authenticity of their message—their vision—is positive, plausible, and palpable, especially if it is one their constituency can believe in.

5.4.4 Detachment and Equanimity

Detachment from an outcome is total acceptance of what is without any desire to have something else, which is a critical concept in true leadership. Detachment, as used here, does *not* mean that one acts without commitment, but quite to the contrary—while one is firmly committed to the principle that serves all people for the greatest good, one is detached from an outcome that would serve only the desires of one's own ego. I (Holly) refer to this characteristic and practice as loving detachment. We are fully present and honestly engaged, but we are personally detached from preconceptions and outcomes in honor of the group and group process.

For example, a person who has worked passionately for a cause may suddenly have the insight that passion placed before principle is a house divided against itself and cannot long stand. Because of this new insight, they now become focused on the principle and, in this process, become detached from their agenda, or the prospective desired result. Thus, to the extent that we can set aside our ego and attachment, we can treat all sides, all points of view, and all possible outcomes with equanimity. Equanimity is the kernel of peace in detachment just as surely as anxiety is the kernel of agitation in attachment.

If a leader is truly detached from the outcome, they will find equanimity to be their touchstone. Equanimity, the outworking of detachment, is reflected in the calm, even-tempered, and serene personality of the one who is simply open to accepting what is. Such a person can lead without wasting energy by reiterating their agenda, their need for or the expectation of approval, or a predetermined outcome. They act out of peace and dedication to the group.

In turn, the peaceful action modelled allows others in the group to relax and see an alternative way of perceiving something, because no one is trying to convince them of anything. They are given ideas and adequate space to consider them. Then, they can ask important questions and, if they so choose, can change their minds in privacy while retaining their dignity intact.

The leader who is detached from the outcome is part of the resolution or transcendence of the problem. Their detachment and equanimity serve to make group participation an exciting process, worth time and commitment, because people feel safe and valued instead of objectified.

5.4.5 A Good Follower

An aspiring leader must learn to follow well before they can learn to lead well. As a leader, you must never ask a group member ("follower") to do something that you yourself are not willing to do—and no one leads forever. It is therefore imperative that you know what it feels like to be a follower—and, as a leader, on the receiving end of yourself.

Consider, for example, a good but insensitive male surgeon who is diagnosed with cancer of the throat and suddenly learns what it feels like to be treated as a patient. As a patient, he must fill out seemingly endless forms and wait interminably to be seen by the doctor. His privileged status as a doctor is moot because not only must he go to a doctor whose specialty is different from his but also his new physician is a woman. And finally, he must face all the uncertainties of the doctor's answers to his frightening questions.

Having seen himself as a doctor through his own eyes as a patient, he subsequently requires his interns to spend time role-playing as patients so they will develop empathy, humility, and understanding of what it feels like to be on the receiving end of a physician's services. This inside experience of being a follower informs the role and quality of a leader.

A good follower must mindfully practice the succeeding characteristics, which are stated in the personal to give you a sense of inner feeling and awareness:

1. Loyalty: A commitment to the group, its process and goals, and its leader and also to the principles of sound leadership. Loyalty manifests itself in your willingness to daily work with enthusiasm on behalf of the group.
2. Understanding: The ability to articulate and integrate into your daily life the vision and principles established by the leader and the group.
3. Candor: The courage to speak your mind clearly, succinctly, and authentically to the group—if necessary, graciously and in private.
4. Active listening: Attention coupled with quieting your mind and observing the subtle nuances of the leader's and peers' communication and behavior if you are to achieve maximum clarity of understanding.
5. Predictability: Being accountable for your own behavior and consistency so the leader and group know they can rely on you.
6. Creativity: Having a beginner's mind that allows you to discover or help discover novel tools and solutions as problems arise.
7. Effectiveness: Getting things done in a manner that helps accomplish the intent of the action.
8. Efficiency: Getting things done in the most expedient and cost-effective manner without cutting corners on long-term goals or compromising either the quality, principles, or effectiveness of sound leadership.
9. Insightfulness: The ability to ask relevant, probing questions and foster innovative ways of seeing and thinking about ordinary things; the ability to advance new perspectives that set the tenor of the success that follows.
10. Honesty: The trait that allows you to understand that the leader and group members can be trusted to accurately represent facts, data, and principles with the highest standard of integrity.
11. Persistence: The tenacity to confront a problem and stay with it until it is either solved or all conceivable possibilities have been exhausted.
12. Communicative: The personal commitment to keep your leader and group members abreast of important developments before they come as surprising news from an external source.
13. Helpfulness: The constant willingness to lend a hand to further the group's vision and/or to uphold the principles agreed upon at the outset.

14. Complementary: The willingness to lend a hand in such a way that your thoughts and actions complement—rather than compete with—those of your leader and peers in achieving a particular end.
15. Equanimity: The choice and determination to maintain an even-tempered disposition; the ability to stay calm and proactive for everyone's sake.[12]

5.4.6 Willingness to Delegate Authority

Delegating authority and responsibility is a vital mark and dynamic of leadership because, when people share the work, more than one person learns the ropes and reaps the satisfaction of a job well done, a victory well earned. Delegating can challenge one's skill as a leader to communicate, listen, plan, make decisions, and solve problems. It teaches one to assess members' acuity and readiness for responsibility, expand rapport, and build productive relationships with all members: those who prefer to follow, those who may like to lead, and those whose expertise is immediately critical to resolving a particular issue.

Revolving authority is the basis of a participatory and pluralistic democratic process—a requirement for a sustainable, self-governing community. Because no one individual can be an expert in everything, particularly in regard to issues involving the complexities of ecology, social, and economic systems and decisions, the official leader must have the good sense and grace to share duties and support and follow the lead of the one whose expertise is in the spotlight. For some, it can be a challenge to seriously share responsibility, prioritize what is best for their community, and step aside to entrust one or more team members.

Leaders must delegate to spread the incremental steps of the project in order to make progress; a time will arise when we must count on someone else's special competence.[13] If we think about the people with whom we share our community, if we understand what a community is at its core: we must rely on one another and look ahead by creating and implementing a collaborative, sustainable vision that meets present and future necessities while protecting our deepest values. By ourselves, we are severely limited, but together we can be something truly awesome. As the proverb says, if you want to go fast, go alone; if you want to go far, go together. Delegating has six basic steps.[14]

These steps are as follows:

1. Choose people who are willing to get the job done and then provide support: advice (when asked), incentives, and motivation. When searching for people to whom to delegate work, it is wise to remember that a person's motivation and dependability are more important than his or her level of skill, which can be learned on the job if necessary.

The way you present a task to someone can bolster their willingness to participate and follow through. It is important for the person to connect this task to their sense of value to the group, the enterprise, and their personal value.

Because each person's values are unique, one must consciously get to know them and actively listen to what they say. Their answers will inevitably reveal their values.

2. Match a particular person to a particular task: This is important because people enjoy using their expertise, especially if they feel the cause is worthwhile, but first, the leader must take time to get to know the person to ensure a proper match. When possible, a good way to help people expand their skills is by asking them about their education, training, profession, family, interests, and/or hobbies.

   If you are organizing a team or committee, it is critical to match people not only to tasks but also to one another because a team or committee must work as a unit in order to be effective. For instance, some people require little or no supervision, whereas others want nothing less than a step-by-step recipe. Some are technicians and like details; others are dreamers and thrive on creativity and global concepts. One person may excel in the limelight, whereas another might be more comfortable in the shadows. Various people prefer to work with numbers, others with words. It is therefore important for a leader to structure a team or committee in such a way that the diversity of personalities, talents, and skills are in a harmonious, complementary working balance.

3. Define the task, communicate it concisely, *and* ask for input. Clearly communicate the purpose of the project to each member of the committee or team and spell out their responsibilities and the project schedule. Specify exactly the desired final outcome, and if necessary, rewrite unclear and/or complicated instructions in the form of an outline or diagram.

   Explain how each task dovetails with, supports, or is contingent on the completion of other tasks in order to integrate them into a creative whole. Then either provide the delegates access to relevant reference materials or tell them how to obtain such materials, if needed. Ask questions to ensure they understand what is expected. Finally, give every member your contact information and be available throughout the project to answer questions and provide additional guidance and/or clarification. Further, provide moral support when barriers occur or things get overwhelming, as they often do.

4. Monitor progress: It is important for a leader to stay apprised of the progress of a project because, without such monitoring, it is easy to encounter inadvertent mistakes through a lack of understanding

instructions or simply human error. Leaders must be positive and inclusive in the way they communicate. For example, "I like the way you have done part A of the task. Do you think it would work to do part B this way? Could that bring it in line with the quality of part A?"

If work is behind schedule or is found to be flawed, it is critical for the leader to remain open and positive. Blame is counterproductive. Blame only shuts people down and puts them on the defensive. One must, therefore, call the team or committee together to discuss the problem and look for such solutions as further clarification of instructions, additional reference materials, restructuring work assignments, and getting more people involved. If it turns out that a particular individual is poorly suited to a task, a mentor can be assigned to work with that person or that person could be reassigned to a task more in line with their capabilities.

If a person must resign because of illness or conflicting priorities and/or responsibilities, empathize with the problem, thank the person for the work completed thus far, and allow them to bow out gracefully. There may be another day or another project on which that person might be willing to serve if today we treat them with gratitude, respect, and kindness.

5. Encourage creativity and allow for different styles of working: A leader must remain focused on the outcome, the final result, not on the detail of how the job gets done. Encouraging creativity infuses a project with interest and vitality. As theoretical physicist John Archibald Wheeler (1911–2008) commented, "In the middle of every difficulty lies opportunity,"[15] which creativity helps to find. Individuals in the group may likely come up with innovative ideas that the leader and other members did not recognize.

   When they are allowed to work independently in their preferred ways, people feel a greater sense of investment and ownership of the project and the outcome. For instance, some people work piecemeal and start with any part of a project, whereas others start at the beginning and complete the whole project with continual effort. Some are dramatic and add a flare to their work; others can't be bothered. Few people probably will work as one expects them to, but they are selected for the quality of their skills and how aptly those skills are applied, however creatively. The way in which the work gets done is immaterial, as long as it gets done on time and is done well and the people had a rewarding time doing it and found personal value in the experience.

6. Always reward effort: A good and sensitive leader is always quick to show appreciation for and recognition of work well done. Such a leader might send a thank-you note as a token of appreciation, hold a

party for all those involved, or write a letter of gratitude for inclusion in the person's personnel file, when appropriate.

An emotionally mature leader always shares the credit and lets people know how valuable their work is. If one treats people well, they are more likely to be available the next time their help is needed, which in turn may allow them to give of their talents and skills in such a way that they find the inner satisfaction of continued personal growth in whatever capacity they choose. Helping others to become good followers is one way a leader can begin to encourage them to develop their potential leadership skills. A positive experience in community collaboration, sharing the credit may also encourage bystanders to volunteer for the next project.

5.4.7 Encourage Leadership in Others

Every leader needs to encourage others to find their own leadership attributes and to help them develop because to succeed in creating and maintaining the sustainability of a community, the community must become, to the greatest extent possible, a community of avid, experienced leaders. All it takes for a leader to encourage leadership in others is to keep eyes and ears open and paper and pencil ready.[16]

One of the most important tasks of a community leader is to continually expand the opportunities for leadership and to support those who have the courage to step forward. As a suggestion, write yourself a short note every time you observe someone in your group assuming the role of leader.

Each person's point of view and thoughtful comments are unique and important. As you help others through your gift of meaningful feedback, you help yourself (the evaluator) to identify things you can use to enhance your own learning and performance in future leadership opportunities.

Further, the sustainability of a community depends on information, and because the meeting of any given committee can be a microcosm of the community "out there," a leader needs feedback as broadly as possible. Once that feedback is available, the leader will have a better idea of how they are affecting the community as a whole and what might be improved. So how can we encourage leadership?

First, nurture and counsel to encourage the leader or future leader. Begin by finding something the leader did well.

Second, be specific. Pointing out explicit behaviors also adds credibility to your comments, which can be thought of as catching someone doing something right and pointing it out to them. To illustrate, I (Chris) was facilitating a visioning process some time back. During the course of explaining how commitment to a vision determines what happens to the community, I paused several times and asked, "Am I making sense to you?" After the process was over, a gentleman remarked, "I really liked how you continually asked us if you were making sense. It not only told me that you really care about us as people and a community but also gave me permission to respond honestly without worrying too much about putting you on the defensive."

Third, suggest an improvement. The last step is to identify a specific behavior that could be improved. Although this may seem difficult, there are several ways to gently suggest improvements. A positive method is to ask a question, such as "Would it work better to make eye contact with the whole audience instead of focusing on one person?" Or begin your comment with "Have you thought of trying . . ." Another way to soften your comment is by turning it into a positive: "Your ability to talk *to* people rather than *at* them is a real gift; it is particularly important that you use this gift during the committee's deliberations about the vision statement."

Being specific and positive when giving feedback is critical, not only because it depersonalizes and neutralizes the observation but because it focuses on a proactive approach and way forward. You could say, for example, "Speaking from your heart, what you really feel, will get your authenticity across to the audience" instead of saying "You need to be more authentic."

Comments must always include a word of encouragement for a specific behavior that is positive and a suggestion for improvement for a particular behavior as needed. Remember, the duty of a true leader is to lead by example, which means keeping your eyes and ears open and your pencil ready for the benefit of other leaders. Now the question becomes how one copes with the demands of leadership.

5.5 References

1. Rod Wagner. 2018. New Evidence the Peter Principle is Real – and What to Do about It. *Forbes* (April 10, 2018). www.forbes.com/sites/roddwagner/2018/04/10/new-evidence-the-peter-principle-is-real-and-what-to-do-about-it/?sh=6cc968751809 (accessed 13 April, 2023).
2. Old Testament, Proverbs 23:7 KJV.
3. Lao-tzu. Tao Te Ching (Book of the Way), Chapter 17, c. 400 BC. Quoted poem in its entirety (Public Domain translation by J. H. McDonald 1996):

   "The best leaders are those the people hardly know exist. The next best is a leader who is loved and praised.
   Next comes the one who is feared.
   The worst one is the leader that is despised.
   If you don't trust the people, they will become untrustworthy.
   The best leaders value their words, and use them sparingly. When she has accomplished her task,
   the people say, 'Amazing: we did it, all by ourselves!' "

4. Daniel Goleman. 2005. *Emotional Intelligence: Why It Can Matter More Than IQ*. Bantam Books, New York, NY., 358 pp.
5. Norbert Wiener. http://members.iinet.net.au/~mazen/quotes.html (accessed February 10, 2022).
6. John Edward Acton. *Quote is Thought to Have Been in a Letter to Bishop Mandell Creighton Regarding Corruption of Rulers, Including Popes*. Archive of Acton's Writings Collected in Acton University, Grand Rapids, MI. www.acton.org/research/lord-acton-quote-archive (accessed January 14, 2023).
7. The foregoing section is based in large part on: Gerald Corey. 1986. *Theory and Practice of Counseling and Psychotherapy* (3rd ed.). Brooks/Cole Publishing Co., Monterey, CA., 406 pp.
8. John C. Holt. 1964. *How Children Fail*. Pitman Publishing Company, New York.
9. Ralph Waldo Emerson. 1875. Social Aims. *Accessible in The Works of Ralph Waldo Emerson*, 8 (Letters and Social Aims) (1909), Public Domain. https://oll.libertyfund.org/title/emerson-the-works-of-ralph-waldo-emerson-vol-8-letters-and-social-aims.
10. Charlotte Perkins Gilman. 1915. Novel, Herland: A Concept is Stronger Than a Fact, Was First Published as a Serial in The Forerunner 1909–1916, and Finally in Book Edition Pantheon Books (1979). Accessible via Project Gutenberg e-book (1992; 2022 update). www.gutenberg.org/files/32/32-h/32-h.htm (accessed January 21, 2023).
11. The discussion of the balance between the masculine and feminine characteristics in a leader is partly based on: Susan St. John. 1997. Making the Emotional Connection. *The Toastmaster*, 63(8):16–18.
12. The foregoing characteristics of a follower are based on: (1) Richard G. Ensman, Jr. 1997. How to Be a Great Follower. *The Toastmaster*, 63(7):19, (2) John S. McCallum. 2013. Followership: The Other Side of Leadership. *Ivey Business Journal*, September/October. https://iveybusinessjournal.com/publication/followership-the-other-side-of-leadership/ (accessed February 14, 2022), and (3) Sandra Slager. *Followership: A Valuable Skill No One Teaches*. Forbes Boston Business Council. www.forbes.com/sites/forbesbostoncouncil/2019/05/16/followership-a-valuable-skill-no-one-teaches/?sh=373d92124695 (accessed February 14, 2022).
13. Frances Moore Lappé and Paul Du Bois. 1997. A Place for Democracy. *YES! A Journal of Positive Futures*, Winter:37–38.
14. The discussion in this section is based on: (1) Judith E. Pearson. 1997. Dare to Delegate. *The Toastmaster*, 63(7):8–11, and (2) Raddha Mehrota. *Delegation for the Win! 6 Steps to Empower and Get More Done*. www.betterup.com/blog/delegation-of-authority (accessed February 14, 2022).
15. John Archibald Wheeler. 1979. The Outsider. *Newsweek* (March 12, 1979). However, not only does the thrust of this quote go farther back in time, it is by now a familiar and ubiquitous axiom of contemporary leadership. See: https://quoteinvestigator.com/2021/10/07/difficulty/.
16. (1) Jean Marsh. 1997. Keep Your Eyes & Ears Open and Your Pen Ready! *The Toastmaster*, 63(1):20–21, and (2) Sydney Finkelstein. 2018. The Best Leaders Are Great Teachers. *Harvard Business Review*, January–February:142–145.

5.6 Discussion Questions

1. What does leadership mean to you? Name some qualities of a leader.
2. If you were a community leader, what would you want to accomplish?
3. Can you think of a true leader in any field, and if you can, what are their leadership attributes that you would wish to emulate?
4. If you had a chance to choose a leader for your community sustainability project, what characteristics would you look for?
5. Have you ever experienced true leadership? How did it make you feel?

6

Coping with the Responsibilities and Pressures of Leadership

6.1 Introduction

This chapter is in part based on circumstances described by Gerald Corey[1] for psychotherapists and my (Chris) experiences in leading workshops on sustainable community development, facilitating vision statements, and mediating the resolution of environmental conflicts in the United States and other countries. As I said in the preface, I believe the circumstances faced by psychotherapists and leaders have much in common. I have also found in leadership that which Corey has found in psychotherapy—namely, leaders soon realize, despite their study and training, that all they really have to work with is themselves—how their own life experiences and values affect their conscience. Their conscience is a beginning point—their wellspring, and also an ongoing touchstone. Leaders must confront the vagaries of circumstances to which they are exposed from their "true north"—from their level of consciousness.[2]

Throughout the intricate dance and process of dialogue and relationship present in every community-sustainability project, the self-aware, thoughtful leader and their leadership acumen evolve in response to the values, ideas, and experiences of the community members, which is a profound give-and-take—a change agent known as two-way learning. In that process, particularly during points of disagreement, difficult choices, or challenges, the leader must again return to and draw strength and direction from their core touchstone. This is a critical concept. The inner charge and purpose of a truly conscious leader is to help the group raise the level of their consciousness, influencing the entire community, such that the world and the way we view our place within living sustainably may be advanced and raised.

DOI: 10.1201/9781003353744-6

6.2 Circumstances Faced by Leaders

A circumstance is a condition or factor that accompanies an event and has some bearing on it. As such, one must consider the circumstance in determining one's course of action to affect a desired outcome. Although no one can directly control a circumstance, each one of us can control how we respond to a circumstance and thereby de facto control it by controlling our response. Self-awareness, reflection, and thus self-control are perhaps the greatest challenges of leadership.

We, therefore, caution aspiring leaders not to be seduced into mimicking or consciously copying another's style, whether it is that of a teacher, mentor, supervisor, or family elder. Principles and practices of contemporary leadership are constantly evolving beyond stale and oftentimes rigid implementation of the past. There is no "right" way to lead, only ethical principles to follow. Although the principles remain the same, there are as many ways to lead as there are people to become leaders, which represents a wide variety of effective approaches to the art of leadership. One way to imagine the art of leadership is as an individual gift that has many facets and must be tended to and developed to fruition. A leader cannot reach all people because not everyone responds equally to the leader's approach, perspective, personality, or level of consciousness.

While an aspiring leader may admire and even be influenced by someone else's individual style, they must choose carefully and be sure to keep clear those boundaries (discussed later in this chapter) that both protect and nurture the gift of their own *unique*, developing style of leadership. Otherwise, the aspiring leader risks becoming an inauthentic, poor imitation of another's approach. Such imitation not only obscures their potential and individual gift but also replicates the other person's limitations or blind spots.

While there is no "formula" for leadership, we know that to lead well a person must work diligently on acknowledging and resolving their own emotional and psychological issues. With this in mind, we can point to a set of the highest, other-centered principles to follow, regardless of temptations to take a route that may be easier, conventional, and self-serving.

As a foundation for the principles, one must locate that place of quiet within, where they are in touch with the heart and their intuition—the knowing beyond knowledge. It is from this place of stillness that an effective leader can move beyond the uncertainty of one's own time to provide support for future community decisions. And it is from this place of tranquility, of detachment from ego, that they may deal effectively with the myriad circumstances to be encountered.

Briefly, our discussion will turn to eight of the many challenging circumstances a leader may encounter: (1) anxiety, (2) criticism in the form of projection, (3) being and disclosing yourself, (4) understanding and respecting the role of silence, (5) the need to be heard, (6) establishing boundaries, (7) learning your limits, and (8) self-deception.

6.2.1 Anxiety

Most budding leaders, and even some seasoned ones, still tremble inwardly with anxiety, which can be called "stage fright" or "performance anxiety," when they anticipate standing in front of an audience or taking their place around the table. They may ask themselves: What will I say? How will I say it? Will the people believe me? Will I make a fool of myself? If I do, how will I ever face them again? Will they accept me or trust me if I make a mistake?

Such anxiety may stem from introversion, shyness, a lack of experience, or a deep sense of insecurity or uncertainty about one's future with those one would lead. After all, the approval of the public at large is portrayed as exceedingly fickle. What, therefore, does one have that is meaningful and essential to contribute? What is one to expect over time? Clearly, one must establish and prove oneself as a leader to lead successfully, which begins with the ability to be assertive.

Being assertive is one of the most challenging skills of effective communication and leadership. Paradoxically, the first step in effectively asserting yourself is acknowledging and owning the anxiety. It took me (Chris) 20 years of engaging an audience of any size—including just one or two people—before I became comfortable being the focal point of other people's attention, in large part because of the way I was conditioned growing up.

The willingness to accept and work with your anxieties, as opposed to denying them through pretenses, is a mark of courage. Although your self-doubt seems normal, how you deal with it is what counts. Returning to the previous theme of loving detachment, practicing meditation is one powerful way to work with anxiety and to attain inner quiet. If you lead from the heart and intuition as well as the intellect, you will find that self-doubt gradually disappears over time until, for the most part, it is no more. As self-doubt fades, your sense of inner, personal authenticity and your comfort (and even enjoyment) in new roles will grow.

6.2.2 Criticism in the Form of Projection

Criticism is really projection, which is casting forward or outward something one perceives to be within oneself, as a way of coping with one's personal discomfort in life. It means the externalization of an inner thought or motive and its subsequent behavior, which is then attributed to someone else. The concept of *projection*, according to the journal *Psychology Today*, originated in a letter written by the founder of psychoanalysis, Sigmund Freud (1856–1939).[3] Projection in the form of negative criticism is a subconscious removal of oneself from personal responsibility. Long before the Freudian concept, we can recognize it in the literature across centuries and cultures:

> [A]nd Aaron shall lay both his hands upon the head of the live goat, and confess over him all the iniquities of the people of Israel, and all their transgressions, and all their sins; and he shall put them upon the head of the goat, and send him away into the wilderness. . . . The goat shall bear all their iniquities upon him to a solitary land; and he shall let the goat go in the wilderness.[4]

In biblical times, on Yom Kippur (the Jewish day of atonement), all the transgressions of the Jewish people were heaped (projected) onto the back of an actual animal called a scapegoat, which was then driven away into the wilderness, "taking" all the people's transgressions with it. The revered anthropologist and philosopher René Girard (1923–2015) elaborated upon this principle and phenomenon at length in his famous theory of scapegoating, also referred to as mimetic theory.[5]

Leaders have probably always been subject to forms of psychological projection, such as criticism. As with many negative behaviors, fear is the root cause. Projection may be especially prevalent in those who would rather reflexively blame rather than take responsibility for their own actions or accept the challenge of examining an event or cause in its more complex context and acknowledging their role. This said, a person who serves the people as a leader must pass the tests described in the eulogy that Senator William Pitt Fessenden of Maine delivered on the death of Senator Foot of Vermont in 1866:

> When, Mr. President, a man becomes a member of this body he cannot even dream of the ordeal to which he cannot fail to be exposed;
>
> of how much courage he must possess to resist the temptations which daily beset him;
>
> of that sensitive shrinking from undeserved censure which he must learn to control;
>
> of the ever-recurring contest between a natural desire for public approbation and a sense of public duty;
>
> of the load of injustice, he must be content to bear, even from those who should be his friends; the imputations of his motives; the sneers and sarcasms of ignorance and malice; all the manifold injuries which partisan or private malignity, disappointed of its objects, may shower upon his unprotected head.
>
> All this, Mr. President, if he would retain his integrity, he must learn to bear unmoved, and walk steadily onward in the path of duty, sustained only by the reflection that time may do him justice, or if not, that after all his individual hopes and aspirations, and even his name among men, should be of little account to him when weighed in the balance against the welfare of a people of whose destiny he is a constituted guardian and defender.[6]

This is the test of leadership: to stand firmly by one's inner convictions even when they are the cause of one's being cast out of office because it was morally necessary to tell the people what they needed to know rather than what they want to hear.

6.2.3 Being and Disclosing Yourself

It takes a great deal of courage to simply be oneself, and not everyone has such courage. Some people become so entrenched in a family, social, professional, or political role that they cannot tell where the role ends and the real person begins.

The other extreme is one who labors hard to demonstrate his or her humanity and, in so doing, overreacts and blurs boundaries between themselves and the members they are leading. If, as leader, one inappropriately discloses too much about oneself (oversharing), for whatever reason, including anxiety or the desire to be liked, they are stealing attention from their constituency and the present mission in an act of self-fulfillment. "Well then," you might ask, "when is self-disclosure appropriate?"

Disclosing persistent feelings that are directly related to the present transaction can be useful, even dutiful. For example, it would be both appropriate and dutiful to express one's observation that community members have not distilled or expressed a clear sense of what they want a particular city park to be like, to therefore help them move forward to clarify their vision of the desired outcome. But regardless of the circumstances, it is wise to always ask these: Why am I revealing myself? What purpose will it serve? To what degree is it appropriate?

6.2.4 Understanding and Respecting the Role of Silence

Many people, including leaders, are uncomfortable with silence and feel compelled to speak. It is not uncommon for a leader to become so threatened by silence that they do something counterproductive to break the silence and thus relieve their anxiety, especially when addressing a small group of people or in a one-on-one conversation. Silence, when allowed to flow unimpeded through indeterminate seconds and minutes, draws people out, causing them to engage in both uncomfortable circumstances and one another.

Just as a painting or other composition of art simultaneously contains defined elements and space, silence is an indispensable, neutral space that is not necessary to fill. Individuals and groups have widely disparate ways of taking in information; consequently, everyone needs time to absorb a message. Some leaders recommend allowing silence for up to a minute or so during a group process before advancing the discussion. When silence feels like a sticking point, a leader would do well to explore its elements. They can be the first to acknowledge the silence, tell the other person(s) how they feel about it, and then pursue the meaning of the silence rather than pretending that it does not exist (for example, by making useless small talk to regain a lost sense of comfort). In pursuing the meaning of silence, one must listen—really listen—to what is being said, because we all need to feel that we are being heard and taken seriously.

6.2.5 The Need to Be Heard

Communication is half giving and half receiving. Listening is the gift of ideas one can only accept through listening. The spoken word that is not heard is like a drop of rain that evaporates before it reaches the Earth.

Active listening features a presence free of distraction. This presence displays compassion with, as opposed to judgment of, another person's sharing of their thoughts, feelings, life situation, and observations. The more a leader demonstrates empathy, the more individuals in the group feel heard, develop a greater bond of trust, and advance understanding among members. Listening requires a quiet, open mind, without the mind-chatter of forming a rebuttal while the other person is still speaking. Genuine listening is an act of love, and anything short of it (whether from impatience or insecurity) is an act of active disregard and disrespect. Such listening requires continual patience over years and is a habit well worth cultivating.

For example, I (Chris) was part of a television program some years ago to discuss the role and processes of ancient forests in the Pacific Northwest. An elderly lady, who also had been invited to be on the program, tried in vain to be heard, but the moderator consistently ignored her. Even after the program was off the air, she tried again to tell the moderator how she was feeling, but he continued to ignore her. In the end, just to be heard, perhaps only by herself, she spoke out loud to no one; she spoke into space. She may as well have been alone in the world.

In contrast, while helping Shinto priests in Japan deal with a forestry issue that was affecting their shrine in Tokyo, I (Chris) sat among 15 to 20 members as they discussed various options. They respectfully listened to me and then to one another, as each priest spoke (in Japanese) quietly, calmly, and without interruptions—even when they disagreed with what was being said. The meeting consumed most of the day, but when it finally came to an end, I was not only informed of the outcome but also asked what I thought about it. That was the most peaceful resolution of a complex issue I have ever observed, and due to the civility, the resulting decision was simply the best and wisest possible.

Not listening is an act of disrespect because it not only interrupts or derails understanding, it intentionally invalidates the feelings—the very existence—of the other. Each of us needs to be heard and validated as human beings. Sharing is the relationship bond that makes us "real" to ourselves, nurtures trust, and gives us meaning in the greater context of our respective communities, society, and the world at large. We simply cannot find meaning without relationship. The trust we nurture becomes built up as positive energy that carries us forward.

Therefore, only when a person has first been validated through listening, can that person really hear what a leader is saying. Only then can a leader share and be informed by another's truth; only then can a leader's gift and harvest of ideas touch receptive ears.

A leader cannot impart their gift if there is no one to receive it—if there is no one to hear. So even if he or she does not ultimately agree, if a leader listens—really listens—and validates the other person's feelings, they can begin to resolve differences among their constituency and thus reduce disputes. The leader's role will be more valuable, much easier, and more impactful short and long terms. Listening involves healthy boundaries; as noted, these boundaries are essential to the foundation of authentic leadership.

6.2.6 Establishing Boundaries

Boundaries are those lines of silent language that allow a person to communicate with others while simultaneously protecting the integrity of their personal space, as well as the personal spaces of those with whom they interact.

The language of boundaries transcends individual space to include familial space, cultural space, and even national space. Understanding personal boundaries among individuals of the same culture is difficult enough but expanding that concept into a fluid working ability among different cultures is the most difficult to accomplish. I (Chris) remember a time in Egypt when I watched a "cultural dance" between the British and Egyptians in a joint meeting. The British maintained a definite, relatively spacious, personal physical boundary, whereas the Egyptians would try to get as close as possible to the British when speaking to them. The British, in turn, would back up, causing the Egyptians to move forward. Thus, continued the "boundary dispute" throughout the meeting, accentuating their cultural difference.

A simple way of looking at boundaries is the adage "good fences make good neighbors." As an example, consider cliff swallows, which attach their mud nests to such surfaces as the faces of cliffs, the sides of buildings, and under bridges. These enclosed, globular nests share common walls, which not only strengthen the nests but also keep the peace by preventing the inhabitants from peeking into each other's abodes. If, however, a hole is made in the common wall and the swallows can see each other, they bicker and squabble until the hole is repaired, which immediately restores tranquility.

How does this concept apply to us in general? Suppose it is Saturday morning, and you leave your home to take care of a few errands. You simply go about your business without paying much attention to what is going on around you or to the people you pass, unless you happen to meet someone you know. In general, you are simply engrossed in what you are doing. When you have finished your errands, you start home.

As you return home, the closer you get to your home, the more you notice what is going on and the more observant and protective you become. Inside your home, the closer you get to your own room, and beyond that to your physical person, which represents your ultimate territory, the more clearly and carefully you define your boundaries.

Although this dynamic may function in a "normal" manner for strangers, it can become so blurred among the members of a dysfunctional family that personal boundaries, including the physical body itself, may be violated. In some families, appropriate personal boundaries are all but absent. This dysfunctional trait is too often carried over into the arena of leadership.

6.2.7 Learning Your Limits

A leader cannot realistically expect to succeed all the time. Even an experienced leader can become glum and begin to doubt their value as a leader when forced to admit that there are people, sometimes whole audiences, whom they cannot reach in a significant way. They must have the self-honesty to acknowledge they cannot work successfully with everyone.

In cultivating leadership, there is a delicate balance between learning your realistic limits and challenging what you sometimes perceive as "limits." You may, for example, tell yourself that you could never work with a certain person or group because you differ so much in your thinking that you could not "identify with them," so there would be a lack of connection or trust.

When confronted by a problem, imagine all the likely scenarios you could use to resolve it. This might include developing a concept map, bubble diagram, or some other visual flowchart of the options open to you. As your skills of simulation deepen, you will gradually become adept at predicting solutions to problems before they occur, which brings to mind a building contractor I (Chris) met some years ago on an airplane.

I asked him how he built a house. "Well," he replied, "I build each house at least a hundred times in my mind before I purchase the first nail or board. That way I see the problems before they arise and have figured out how to fix them with the least cost and loss of time. I'm the only person who really 'knows' the house. I know it better than the owners ever will, even if they live in it for 50 years!"

However much help you might think you need to expand your perceived limitations of leadership, you undoubtedly have talents and skills that you can offer to someone else. Do it. Act as a mentor to a less experienced peer, and you will find that you are also helping yourself by sharpening your own skills of human relationships that will, in turn, motivate you to learn and excel in your own ability to lead while helping to avoid self-deception.

6.2.8 Self-Deception

Self-deception is not consciously lying to oneself; it can be subtle and unconscious. A leader's motivation for deception is the need to make the relationship with followers both worthwhile and productive. A person's need to see change in the form of forward movement may blur reality and cause them to be less discerning than is wise because their role as a leader carries with it an investment of succeeding as a leader and, conversely, the risk of failure.

What would happen if one had the sense they were a flop as a leader? A leader's need to feel instrumental in assisting others is partly based on one's own need to make a significant difference in the world, a need that as ego-attachment can, and does, advance self-deception.

At such times, a leader may look for evidence of progress, however slim, and rationalize away any elements of failure. In so doing, they may claim credit for the apparent progress when it may be largely due to the group as a whole, as well as other factors unrelated to their leadership. When a leader understands that success or failure is not the event in and of itself but rather their interpretation of the event, they can consciously explore the phenomenon of self-deception. Learning the value of failure, as well as how to fail, is a practical skill that acknowledges the realities of visioning, leadership, and decision-making. Another value in detaching from outcomes (loving detachment) allows us to accept failure as an irrefutable ingredient in all processes, whether individual or group-oriented.

6.3 Leadership within Organizations

Organizations, including governments and corporations, cannot empower people, but they can give them the proper tools and the capacity to work actively in achieving the positive outcomes they want. For instance, incentives for good trusteeship of a local community's natural resources by the community itself can be put in place as a priority over possible tax breaks to recruit multinational corporations with little or no stake in either the ecological sustainability of a community's natural resources or the high-quality jobs needed to sustain a community's economic and cultural well-being. To illustrate, a community's leaders could create a trained workforce of community members to do such things as maintain the sustainability of a wetland, existing open spaces, and/or other biophysical projects that would benefit the community's overall social-environmental well-being. The money would thus be spent locally in the community's favor instead of being exported as a tax break to some outside corporation with no real commitment to the community's long-term sustainability.

In addition, organizations for economic development can provide thoughtfully tailored expertise and financial backing, rather than generically ladling out grants, so that local community partnerships can learn how to create their own vision, strategic plans, and steps toward self-sufficiency to achieve their goals. Such support would include assistance in research, facilitated processes, strategic planning for businesses, and marketing, to help communities retain and attract bedrock small businesses critical to their social sustainability.

Local leadership can also employ performance measures for agencies and their staff to reward those actions that help the growth and sustainability of community partnerships. In this spirit, a mayor, city or county council, or governor can require the use of cross-agency or cross-departmental decision-making and can combine funding for programs in order to increase the efficient and effective access of local communities to resources at the agency's or department's disposal. Such collaborative structures are a form of ecosystem-based management at two levels: (1) horizontal (coordination of organizations at commensurate levels) and (2) vertical (coordination of organizations from local grassroots to national, regional, and international levels).

Imagine, for example, cooperation and coordination among water conservation goals in geographically large river basins, including the connectivity of habitat corridors for the seasonal migration of fish, birds, and mammals. Greater teamwork can help achieve larger, more holistic goals both on land and at sea, for example connectivity among the habitats within marine protected areas.[7] Such coordination can save money and personnel and gain traction among otherwise competing groups. Increased devolution of decision-making authority to the local level with active participation and engagement can generate lasting, innovative community partnerships and thus more effective results, which creates the opportunity for community self-empowerment.

A prerequisite for sustainable development in a local community is that it must be inclusive, relating all relevant disciplines and special professions from all walks of life. Setting a good example is one of the most important functions of any local organization or government involved in implementing the principles and practices of community sustainability. Leading by example—breaking down bureaucratic barriers of turf through the interdisciplinary crossing of departmental lines, recycling and buying recycled goods, providing day care, encouraging carpooling, and offering flexible working hours—increases not only the capacity of an organization to govern its own people but also that organization's effectiveness *and* efficiency. The social-environmental sustainability project that your community tackles may be a positive example capable of being adapted to other communities' advances in the future. Capturing the process and its landmarks will help your project live beyond its own boundaries.

In linking organizations to multiply benefits, it is important for them to both identify departmental and community entities or stakeholders with mutually interrelated issues and to bring all who are affected to the discussion table. Any effort to collectively resolve shared problems requires embracing and working creatively with community diversity. Welcoming diversity allows us to acknowledge that each of us has something special to contribute. Moreover, the need to be accepted and included is the essence of being human. The greater the involvement with one's community, the greater the tendency to want to stay engaged and participate in future initiatives. It also enables us to recognize that no one has all of the answers, that we do not—and cannot—know or do everything, that we are interdependent, and that we must welcome, trust, and rely on the strengths of others.

Diversity of thought, culture, expertise, and economic status allows all persons to contribute to the process of their community's social-environmental sustainability. Conversely, just as simplifying an ecosystem or complicating a mechanical system increases its vulnerability to destruction, so, too, will segregating diverse elements within a community lead to isolation, and the community's social, moral, and economic atrophy and deterioration.

Once the group embraces its diversity, what do they most want from one another? People want the most effective, productive, and rewarding way of working together to achieve a common end. They want the process and the relationships forged to meet their personal desires for belonging, meaningful contribution, and commitment to a special place (their community), the opportunity for personal growth, and the ability to exert reasonable control over their own destinies.

Control over personal destinies and, thus, the destiny of a community can be increased if public agencies will focus on and support a community's vision and goals for social-environmental sustainability. Having said this, however, it must be recognized that federal agencies (*not* a local community per se) have jurisdiction over state and federal public lands and thus (as stewards, under the ancient public trust doctrine[8]) are ultimately responsible and accountable for the ecological sustainability of those lands, as a national legacy for all generations—present and future.

It is also possible to create organization-to-community trusteeship contracts—including the government, local and otherwise—that recognize it is the results that count. Such action can honor and support community visioning and simultaneously allow employees at all levels of government to be part of the vision, to empower themselves to achieve quality results, which may well improve ongoing and future citizen participation in self-governance and the community's social-environmental-sustainability vision processes.

If we, as adults, do nothing else, we can and must find the moral courage to lead by example in how we treat one another and thereby help teach our children the absolute social necessity of common civility. Consider that all we have as real earthly value is one another—someone with whom to share the experience of being human, with all its frailties, uncertainties, and magnificence. We *need* one another to be whole and healthy. That said, the painstaking building of community and connectedness, recognizing and actively participating in our interdependence, takes time, patience, and courage, yet these must be our centerpiece priority for sustainable progress, our planetary stewardship, our full humanity, and our survival. One cannot share in life's experiences and cannot offer the gift of one's own experience and talent if there is no one to accept it. As Catholic nun, human rights advocate, and Nobel Peace Prize recipient Mother Teresa (Mary Teresa Bojaxhlu, 1910–1997) said, "Loneliness and the feeling of being unwanted is the most terrible poverty."[9]

6.4 References

1. Gerald Corey. 1986. *Theory and Practice of Counseling and Psychotherapy* (3rd ed.). Brooks/Cole Publishing Co., Monterey, CA., 406 pp.
2. Ibid.
3. According to *Psychology Today*, Projection, Sigmund Freud first discussed the concept of projection in a letter in 1895; the concept continues to be debated and developed. See: www.psychologytoday.com/us/basics/projection (accessed January 21, 2023).
4. The Old Testament KJV. *Book of Leviticus*, 16:21–22.
5. René Girard. 1966. *Deceit, Desire and the Novel: Self and Other in Literary Structure.* The Johns Hopkins Press, February 1, 1966. ISBN-10: 0801802202, ISBN-13: 978–0801802201.
6. (1) John F. Kennedy. 1961. *Profiles in Courage*. Harper & Row, New York, NY., 266 pp. and (2) *Congressional Record*, 151(69): S5807S5813. www.congress.gov/congressional-record/volume-168/issue-151/extensions-of-remarks-section (accessed February 15, 2022).
7. Chris Maser and James R. Sedell. 1994. *From the Forest to the Sea: The Ecology of Wood in Streams, Rivers, Estuaries, and Oceans.* St. Lucie Press, Delray Beach, FL. 200 pp.
8. (1) Joseph L. Sax. 1970. The Public Trust Doctrine in Natural Resources Law: Effective Judicial Intervention. *Michigan Law Review*, 68(3):471 and (2) J. B. Ruhl and Thomas A. J. McGinn. 2020. The Roman Public Trust Doctrine: What Was It, and Does It Support an Atmospheric Trust? *Ecology Law Quarterly*, 47:117–178.
9. Mother Teresa. 1975. *Time* (December 29, 1975).

6.5 Discussion Questions

1. As a leader, what might be the most daunting circumstance you could possibly confront in your community?
2. What do you think would be your highest leadership responsibility to your community?

3. What sort of boundaries would you have to establish as a leader in your community?
4. How might you constructively deal with occasional criticism?
5. What are some ways you could grow in a leadership role? What are some ways you could empower your community members to also grow in their leadership capacity?

7

Negotiating Constraints to Community Visions

7.1 Introduction

True progress toward an ecologically sound environment and an equitable world society may seem expensive in terms of cost and effort. However, the costs of a degraded environment (land, air, and water) are far higher in terms of damages and lost ecosystem services that are rarely valued in economic terms. The longer we wait, the more difficult and expensive the restoration (repair) of biophysical-system processes, the more elusive the success of conservation, and the greater the damage to the system: environment, people and their communities, livelihoods, public health, and the overall economy.

Biological shortcuts, technological quick fixes, or political promises cannot ultimately mend what is broken. Dramatic, fundamental change from the bottom up is necessary if we are truly committed to the world's children, present and future. It is not a physical question of whether we can change but rather one of whether we will embrace change and *commit* to it. Whatever our decisions, we, the adults of the world, bequeath the next generation and beyond the consequences of our actions as members of our democratic, participatory governance. *We are all responsible for our decisions, whether or not we choose to participate consciously by staying well informed and actively voting.*

Insisting on little tweaks to address problematic symptoms—but not the problem at its source—inevitably yields fragmentary results without regard to the larger context of the social-environmental system itself. In other words, immediate, seemingly easy fixes do not put us on target to effectively solve our interrelated problems.[1]

What society really needs are "big fixes" in the form of ideas stemming from interdisciplinary systems thinking that promote and safeguard social-environmental sustainability while understanding that any "fix" must be true and realistic in terms of its biophysical foundation and must be continually adapted or upgraded based on experience and evolving knowledge.

DOI: 10.1201/9781003353744-7

Where, asked the late publisher Robert Rodale, are the "big ideas," those that change the world? Likely, they exist but remain unrecognized because our quick-fix, symptomatic social trance does not allow the time, patience, or intellectual freedom and breadth of understanding necessary for true, systemic solutions.

A big idea, according to Rodale, must

1. be generally useful in good ways—leadership that is multidimensional in its decision-making and translates into social-environment sustainability, as opposed to unidimensional decision-making, which is strictly focused on a position to defend, be it a corporate bottom line or a military objective;
2. appeal to generalists and give them a leadership advantage over specialists—true social-environmental decision-making requires an understanding of the system as a whole *and* so necessitates an amalgamation of broad, systemic-oriented thinkers *and* narrow, more symptomatic-oriented thinkers—with systemic thinkers in charge;
3. exist in both an abstract and a practical sense—decision-making, as seen in item 1, not only is practical in its outcome but also abstract in that its practical outcome requires people to work together toward the conceptualized ideal of a shared vision with love, respect, humility, wonder, and intuition, as well as their intellect;
4. be of some interest at all levels of human concern—social-environmental sustainability requires the continual building of relationships, which requires sensitive, systemic-thinking leaders who make wise decisions because they touch all levels of society, both within itself and with nature;
5. be geographically and culturally viable over extensive areas—systemic-thinking leaders are a general necessity if the natural world is to remain viable and habitable for future generations;
6. encompass a multitude of academic disciplines—leaders promoting systems thinking require collaboration with various specialists since every complex issue requires insights across the disciplines, such as soil sciences, mycology, philosophy, sociology, ecology, forestry, economics, religion, education, governance, and other sciences and fields to make the wisest decisions for the long-term social-environmental sustainability; and
7. have a life over an extended period. Social-environmental sustainability is, by definition, a journey of increasingly well-informed decision-making across generations. This decision-making requires a succession of skilled, systems-thinking leaders who know how to facilitate and guide a shared vision toward fruition, as experience accumulates and new knowledge emerges.[2]

Multidimensional leadership that fits Rodale's criteria can help people understand that life is not reducible to a black box: any model is necessarily a shorthand representation, operational simplification, or working hypothesis in need of improvement—most often through empirical, on-the-ground data. When we accept that there are neither shortcuts nor concrete facts other than direct observation (such as glaciers melting worldwide), we will see how communication functions as a connective tool through which we can and must share experience, invention, cooperation, and coordination in all aspects of social-environmental decision-making.

To protect the best of what we have in the present for the present and the future, we must continually broaden and inform our understanding and behavior to accommodate new circumstances, knowledge, and conditions. Society's saving grace is that we are resilient and have choices. As stated earlier, the goal is to consciously protect the range of choices available to future generations. Accordingly, we can achieve what needs to be done by being well-informed, other-centered, determined, and *committed.*

7.2 Identifying and Understanding Constraints

When people speak from and listen with their hearts, they unite and produce tremendous power to formulate new realities and bring them into being through collective decisions and actions. Therefore, a thoughtfully crafted community project for sustainability that recognizes and honors the productive capacity of both people and ecosystems (within their particular biophysical principles) is critically important. After all, social-environmental sustainability is a long-term path and in many ways an enlightened course correction—our choice, but one that must be carefully planned with humility.

While seemingly simple, humility is a deceptively complex concept. In the planning context, it entails being open to learning new things, listening to new ideas, practicing imaginative "big" thinking, being honest but respectful, making mistakes and getting things wrong sometimes (and backing up or starting over), and having a commitment to patient, ongoing collaboration with fellow community members (some or many who may be strangers at first). Humility requires us to let go of attachment to fixed ideas or prejudices, mental models, preconceived notions, ego, and pride. All of this takes courage. At the end of the process, we are different people, because we have grown individually, as a community, and we have built something together that sustains us for the present and future. It becomes for us as a famous Tibetan Proverb states: "In deepest conversation, you and I become we and us."[3]

7.3 The Community of Tomorrow: Where Do We Want to Go from Here?

7.3.1 Mistaking the Map for the Territory

While around the world we might have much in common, and our goals and objectives might be similar, each community is unique. Similarly, no two community-sustainability projects are alike. The reasons for the differences in projects may seem self-evident and include geology and geography, culture and values, climate and precipitation/water availability, accessibility to land and water rights, planning expertise, funding, and the scope and scale of the project. The main point is that, unlike a house plan selected out of an architectural pattern book, it is neither realistic nor possible to precisely duplicate, overlay, or transfer a project plan from one location to another. Adopting this understanding from the beginning encourages an appreciation and open-minded engagement with what "is," in terms of the factors present that not only support but also constrain project outcomes. In other words, we make uniqueness an asset that directs our design.

For example, perhaps a desert community is planning a green space with native vegetation that provides cooling, forage for birds and other wildlife, some benches for resting and enjoying the surroundings, a gravel walking or bike path, or other functional and aesthetic amenities. One community member has suggested a water feature, such as a small fountain or a waterfall with native rock. The water feature would further enhance humidity and cooling and generate ambient sound that is pleasing and relaxing. Using precious groundwater or water from a small stream might strike some as a poor use (or even a waste) of water. However, suppose in this particular scenario there is the chance to discuss with local municipal and business leaders an aspect of the project that would include creating co-benefits in collaboration with adjacent uses. Examples of co-benefits could include restoring a small wetland onsite, filtering and recycling landscaping water from the nearby golf course or stormwater runoff from the adjacent built environment resulting in the reuse (and therefore conservation) of water in an arid environment. Co-benefits promote efficiency because they effectively integrate efforts to multiply and amplify results; co-benefits save money by combining purposes into one project.

Whether we are talking about a project in the desert, the tropics, or the far north, every scenario presents constraints due to each site's unique geography, climate, precipitation, population density, surrounding uses, and so forth. These are called biophysical constraints. Additional constraints can include (1) budget; (2) materials, such as lumber and concrete (the price-spikes and shortages of which—or long wait for—many communities experienced during the 2020 pandemic); (3) personnel and expertise available for advising and consulting on the planning; (4) project time frame; (5) local land-use and zoning guidelines and requirements; and importantly, (6) adequate time to gather public opinion or input on the options for the plan.

7.3.2 Being Realistic: Taking Scope, Scale, and Time into Account in Planning

By taking our time and troubleshooting in the beginning, some constraints can be foreseen and thus avoided. In other words, thinking ahead (visualizing and analyzing the parts and the whole) prevents common mistakes. Just as if we were planning a project at home, for school, or for work, selecting a realistic time frame from start to completion must be informed by honestly appraising several factors: (1) How complicated is the project? (2) How large is the project? (3) What materials and supplies are necessary, and are they available and easily obtainable? (4) What are cost estimates based on research? (5) Who needs to be involved and are they available? (6) What are the individual time requirements necessary for each unit or building block of the overall project?

Conservative planning means proactively building in extra time for unexpected delays. Still, every time frame is only a rough estimate using the most accurate input and best judgment we have from the beginning. Are there neighbors, peers, experts, and leaders in another city or county we can consult about their recent, roughly similar project (such as the earlier example regarding the green space or park)? Every project has a story, and ideally the stories of sustainable community initiatives need to be gathered and shared with the wider regional community and beyond. A major reason such stories need to be imparted internationally is that we are presently seeking new ideas, methods, and concrete examples of community sustainability based on efforts and successfully finished products because we are undertaking the greatest work of our age. We are reinventing the way we live because our everyday assumptions from an earlier historical period are no longer to be counted on (if ever they were to be trusted) for the simple reason of their being prior assumptions.

7.4 Attributing (or Misattributing) Causes and Correlations

To introduce this section, I (Holly) will share a story from the 1941 nonfiction *The Log From the Sea of Cortez* (by marine scientist Edward F. Ricketts [1897–1948] and writer and recipient of the 1962 Nobel Prize for Literature John Steinbeck [1902–1968]). In *The Log*, the authors describe their cruise on a retrofitted purse-seiner from Monterey, California, south to the Sea of Cortez (the Gulf of California between the Baja Peninsula and Mexico).[4] In the book, after the authors delve into the major theme of the difference between teleological (might be, could be, belief, and emotional) and nonteleological (is) thinking, they impart an important anecdote concerning the desire by Norway to protect the endangered willow grouse, a popular game bird. Using visual observation led researchers to determine that the cause of the birds' decline was a major predator, a hawk. The region, therefore, put a bounty on the hawk, to no avail. The grouse declined even more precipitously until the researchers broadened their investigation.

> An ecological analysis into the relational aspects of the situation disclosed that a parasitic disease, coccidiosis, was epizootic among the grouse. In its incipient stages, this disease so reduced the flying speed of the grouse that the mildly ill individuals became easy prey for the hawks. In living largely off the slightly ill birds, the hawks prevented them from developing the disease in its full intensity and so spreading it more widely and quickly to otherwise healthy fowl. Thus the presumed enemies of the grouse, by controlling the epizootic aspects of the disease, proved to be friends in disguise.[5]

There is always more to a situation than meets the eye, whether in the environment or within society. Several factors, rather than one, often influence a phenomenon. As the popular quip goes, "it's complicated!" The antidote to automatically grasping for immediate answers is to (sometimes tirelessly, as in science and medicine) ask questions including why and how something occurs—avoiding the tendency toward snap judgments.

The need for proactive troubleshooting applies throughout the steps of every sustainability initiative; equally important, it allows an estimate of the initiative's required time frame for completion. For example, troubleshooting includes understanding the dynamics of both the biophysical and social constraints, many of which are invisible to the naked eye. Both the environmental system and the social system are deeply complex, and the more complex a system is, the less valid the use of assumptions is, because what we observe in a biophysical system is only part of the total picture.

To illustrate, imagine a community project for the renovation of a very large parcel of riverside land that the public has purchased from the original property owner. The beginning idea for the new, multi-year initiative involves the demolition of abandoned industrial sites, reconnection to a historic floodplain, creation of public gardens in keeping with a riparian zone, creation of ponds connecting surface to groundwater and providing flood attenuation, soil testing, contaminant remediation and restoration of soils, a public green way or park, and low-impact development consisting of various economic levels of housing, as well as a retail store and a restaurant. Parking and walkway areas will all be constructed of primarily permeable surfaces that filter precipitation instead of generating sheet flow (runoff, such as from asphalt or concrete after heavy or prolonged rains). The landscaping prioritizes trees and other plantings that are native and provide shade, pollinator and wildlife habitat, and other ecosystem services.

Without historic photographs (if they exist) and contemporary technological tools, we would have no idea how to locate the historic river channels or floodplain—remnant traces of which are not visible to us. Similarly, without laboratory testing, we would have no idea whether the area's soil and water are contaminated and require our attention (although perhaps we could guess). For example, was the prior land dedicated to building ships with asbestos and other materials or manufacturing batteries with lead, copper, or cadmium? In addition to information about specialized prior use, we are concerned with other, commonly found urban contaminants from a long-occupied site, including radon, asbestos, creosote, petroleum residues, fertilizers, mercury, pesticides and herbicides. In addition, certain contaminants, such as arsenic, are natural in some soils around the world.

So embedded in our illustration are at least two invisible factors that must be further investigated with the support of science and relevant experts: (1) the historical dynamics of the floodplain and (2) the quality of the soil and water composing the riverside land parcel in general. Gathering as much data as possible before planning commences prevents costly problems later, including public exposure to toxins.

Before demolition and remediation, we may notice water on the site pooling after a storm or large patches of ground where nothing grows—not even weeds. These are clues that could be easily misattributed without further detective work; we might, for example, chalk these areas up to the ground being compacted after decades of heavy machinery, then fail to look more closely at causes (asking why and how). A closer examination might reveal that the soil has no aroma or else has a repellant odor (good soil smells rich and pleasant). Examining an onsite soil sample under a hand lens or microscope reveals no intact, biodiverse soil features, such as an abundant community of interdependent life forms—but rather no life whatsoever. By looking more closely and asking good questions, we will have gathered important clues about our site, clues that were previously invisible but are relevant in helping us build a successful project.

With the foregoing in mind, it is critical to understand that, beyond biophysical constraints and those of time, scope, and scale, we must be able to work together effectively (efficiently and harmoniously). Previous chapters have described the ingredients of powerful, cohesive group dynamics. In fact, many concepts and themes throughout the book apply directly to the functionality of community projects based on sustainability. To summarize, these concepts and themes include respect, selflessness, willingness to creatively engage and work through conflicts or disagreements, commitment to the process, and inclusive and attentive decision-making, to be extensively discussed in the next chapter, Chapter 8.

7.5 References

1. Chris Maser. 1992. Do We Owe Anything to the Future? In: *Multiple Use and Sustained Yield: Changing Philosophies for Federal Land Management?* Proceedings and summary of a workshop convened on March 5 and 6, 1992, Washington, DC., pp. 195–213. Congressional Research Service, Library of Congress. Committee Print No. 11. US Government Printing Office, Washington, DC.
2. Robert Rodale. 1988. *Big New Ideas—Where Are They Today?* Unpublished speech given at the Third National Science, Technology, Society (STS) Conference, February 5–7, 1988. Arlington, VA.
3. Tenzin Gyatso. 2009. *The Art of Happiness, the XIVth Dalai Lama*. Riverhead Books, New York.
4. John Steinbeck and Edward F. Ricketts. 1941. *Sea. Of Cortez: A Leisurely Journal of Travel and Research*. First published in the United States by The Viking Press—now owned by Penguin Random House, New York, NY., 598 pages.
5. John Steinbeck and Edward F. Ricketts. 1995. *Sea of Cortez: A Leisurely Journal of Travel and Research*. Penguin Books, New York, NY., Page 120.

7.6 Discussion Questions

1. Many community sustainability planning processes and projects take years. How might you support your fellow community members to maintain energy and focus?
2. What roles do efficiency and effectiveness play in community planning, and is it important? How and why, if so?
3. In building capacity for community members to participate in planning, what are some of your ideas for your toolbox: specifically, methods of effectively teaching citizens proactive ways to engage with planning constraints, such as delays, budget issues, or disagreements?
4. Does the uniqueness of each community pose social-environmental constraints to the outcome of its vision for the sustainability of its future? Think of some examples.
5. What do you see as potential constraints in the planning process in your community?

8

If You Want to Go Far, Go Together by Making Inclusive and Intelligent Decisions

8.1 Introduction

This chapter builds on concepts involved in making planning decisions in partnership with the unique characteristics of the land and a community's vision by elaborating on individual involvement within the circle of fellow community members. The chapter emphasizes the importance of inclusivity, collaboration, and improving the group process through sharing, listening, and learning from the beginning to the completion of the project. The chapter concludes with a summary of the ingredients of sound decision-making.

8.2 Inclusive Brainstorming for Group Success

A major, preliminary part of any plan is choosing the participants who must be involved. This selection depends on the size, scope, and details of the vision; project location and description; potential users of the space involved; and the planning process itself. The general answer to "Who should be involved?" is *as many people as possible*, to arrive at a group that fully and fairly represents a cross-section of the community.

In the end, the members of the group will be self-selecting. However, it is crucial in the beginning to alert everyone about the project so no one might feel left out. After all, the more representative and diverse the group's membership, the more well-rounded and well-informed the input, ideas, suggestions, options analyzed, and the decision process will be as a whole. When residents choose to get involved and learn, the outcome more accurately reflects their dreams, priorities, and values, as well as those of their neighbors. Being involved is the heart of being a community; participation in shaping the future is our birthright.

DOI: 10.1201/9781003353744-8

The United Nations General Assembly recently voted on a Declaration of Human Rights (2022, an evolution of the 1948 Declaration), which states that all people have a universal human right to "a clean, healthy, and sustainable environment."[1] *Note:* The right to a sustainable environment is recognized as equal with all our other specific human rights (enforceable in the countries that signed on to the Declaration, which overwhelmingly passed). The recognition of our urgent—yet ancient—global connection to the practice of environmental sustainability, as it relates to our human health and well-being (including overall human equality) is still evolving throughout every aspect of society, as we find and practice ways to expand it on the ground, and in both our systems of governance and economics. Thus, connecting our communities to opportunities for participating in the development of sustainable places, practices, products, and the ongoing reimagination and replacement of outmoded ways of living can help ensure the human dignity and success of the generations who follow.

The conceptualization, discussion, and implementation of groundbreaking ideas require sharing diverse perspectives from those who wish to shape the vision of social-environmental sustainability *now* for those in both the present and the future, but the process and implementation of envisioning change also require partnerships among community members in general and those who have the necessary expertise and experience, funding, leadership, and official mission to support our efforts. These partners play vital roles in every community project related to sustainability. The defined partners and roles are unique to each vision. Therefore, specific choices of partners for each initiative of community sustainable will depend on the goals, project description, and who is available and willing to help.

As emphasized previously, nature is the first and foremost partner; we must make decisions in partnership and keeping with the relevant surroundings—the physical environmental systems' biophysical characteristics and limitations, as well as what we already know about how change is evolving (lower rainfall, for example). In addition to our core group of community members who want to shape the project, we must consider including whoever can help us make it an enduring success.

8.3 Partners and Their Roles in Community Planning

The following is a list of potential partnerships that community groups typically include in their planning projects and examples of the roles they play.

While a particular project may require just a few of these roles for sustainability, the list is provided as a touchstone regarding discussion:

1. *Architects*, including landscape architects
2. *Biologists*, who can help provide detailed data on the site, including species of plants, animals, insects, and their habitat requirements
3. *Community leaders* in relevant capacities, such as land-use zoning, permitting, and planning staff; local parks, recreation, and greenspace officials; nonprofit environmental organizations; local businesses; and local, state, regional, and federal agency staff
4. *Contractors and builders* of various expertise, for such things as a community center, playground, sports field, picnic area, fountain, arts stage, or classrooms
5. *Designers*
6. *Educators*, including recreation specialists
7. *Engineers* there are five branches in the United States, one of which is civil engineers—a group that includes biological and ecological engineers, who can survey the site to determine elevation, and traffic engineers for help with designing safe access to and exit from the area
8. *Facilitators* trained in supporting group focus, dialogue, and decision-making
9. *Fire* and other public safety experts and representatives
10. *Foresters and forestry experts*
11. *Geographic information system experts* (GIS, for mapping)
12. *Horticulturists*—experts in native plants and others suitable for the area, the design of an edible garden, or a children's garden—and other scientists
13. *Hydrologists and geologists*—experts in how surface and groundwater interact with relevant landscape features and how to incorporate permeable features that attenuate or prevent flooding—instead of impermeable surfaces that exacerbate flooding, such as concrete or tarmac
14. *Landscape architects*—to maintain a beautiful, sustainable community panorama
15. *Local businesses*—to maintain a sustainable economy within the community
16. *Local/state/regional/federal agencies*—responsible for maintaining the various laws governing citizen behavior with respect to the communities social/environmental long-term sustainability

17. *Project funders and grant makers*—examples are vast and can overlap with government entities, businesses, charitable foundations, and nonprofit organizations
18. *Stakeholders*—everyone who will be affected by the outcome of the decisions made
19. *Transportation experts*—enhancing accessibility via public transportation while protecting the connectivity of the landscape

This is but a small selection of suggestions. The unique goals, requirements, and characteristics of each sustainability initiative will dictate relevant categories of support.

8.4 Informing Projects with Relevant, Local Science

Whether or not a project site has already been selected, gathering existing (and even historic) information is important. Existing information can be located in county surveyors' offices; courthouse documents, such as deeds; local university (and other public) libraries; and state or regional offices. To augment existing information, it is beneficial to identify what additional, contemporary types of studies would provide crucial data to inform our decisions on the land. Taken together, these analyses (both old and new) provide insightful information for the planning group's discussion and understanding of the specific biophysical features (and limits) present. In other words, what is possible, practical and realistic, achievable, and sustainable for our planning purposes—which proposals are harmonious and work *with* the land.

Gathering existing information, identifying potential experts, and ordering and arranging new studies take time. New studies may require detailed bids and contracting processes. The main point is that, in the budget and the timeline, there must be itemized categories for gathering and organizing data. Some studies are best done during certain seasons, so they must be planned accordingly.

8.4.1 Citizen Involvement in Science for Planning

Are there opportunities for everyday people to work with a team leader to make relevant observations, contribute, and record data? The answer, which is yes, describes citizen science:

> Around the world ordinary people of all ages engage in citizen science—participating in projects in which volunteers and scientists

> work together to answer real-world questions. Much of this work is conducted close to home, sometimes in our own backyards or even in our living rooms and kitchens, with guidance from professional scientists and using established science protocols and tools. Regardless of the location and process, citizen science brings everyone into the important work of learning more about and protecting our planet.[2]

In addition to contributing to community sustainability, we can also add to information about species in our area through actions as simple as taking a photograph with a cellphone camera; for example, a photograph of a plant that is uploaded to an application, such as iNaturalist or Plant.net, helps to identify species and contribute a sighting to the database. Such observation can also help us accurately identify trees, birds, butterflies, and more out in the field and learn about them in our area.

Citizen science is complementary to outdoor education programs, kindergarten through high school, and out-of-school adult education and helps us provide inspiration and hands-on learning experience for young people and their early, formative development in science, technology, engineering, and math (or STEM). In one estimate, almost two million people worldwide participate in citizen science on an ongoing basis.[3]

8.4.2 Maintaining Flexibility in Thinking *and* Planning

As mentioned throughout the book, change is the only constant process. It is, therefore, natural for community visioning and sustainability projects to encounter challenges and temporary setbacks. While many of these can be identified and proactively prevented early on through troubleshooting, challenges are inevitable. Embracing occasional disruptions in planning can not only help us practice resilience but also help us support the group's journey and successful completion of the initiative or project. The old saying that "nothing worthwhile was ever easily won" stands as a good reminder that the greater our ambition, the more work—particularly teamwork—the realization will require. Along the way, there will be hundreds of small and large decisions, all of which affect the undertaking in various degrees.

8.5 A Proposed Set of Decision-Making Guidelines

Just as there are inviolable biophysical principles that govern nature and the universe, so there are inviolable guidelines for decision-making. Here, the term inviolable is used advisedly; should one attempt to cheat in applying nature's principles through poor decision-making, the costs will be borne by everyone for a long time. The problems plaguing the globe today are the products of our self-centered thinking based on our own perceived needs and short-term profit motives.

To change anything, we must reach beyond where we are, beyond our well-worn, reflexive comfort zone. We must dare to move ahead, even if we do not fully understand where we are going, because we will never have perfect knowledge. Thus, we must ask probing, innovative, other-centered, future-oriented questions to make the necessary changes. As we take on community initiatives to enhance sustainability, it is essential to elaborate on these leadership guidelines.

8.5.1 Guideline 1: Everything Is a Relationship

8.5.1.1 Intra-personal

An intra-personal relationship is an individual's inner sense of self-worth, personal growth, authenticity, and so on. In short, it is the degree of psychological maturity that makes a person conscious of and accountable for his or her own behavior and its consequences. The more conscious we are, the more other-centered we are, the more self-controlled our behavior, and the greater our willingness to be personally accountable for the outcomes of our behavior with respect to the welfare of fellow citizens, present and future, and Earth as a whole.

People who succeed in changing their lives and growing beyond their present limitations of consciousness do so because they not only know that success is the interpretation of an event and not the event itself but also because they recognize their own inner patience with themselves and thus self-control, which gives them a sense of personal authority.

Each decision is a fork in the road; each presents at least two choices—right-hand fork or left-hand fork (but frequently far more choices). Whichever fork you take, you get everything it has to offer and forgo whatever the fork not taken has to offer. The direction of our lives is determined by the cumulative results of many little decisions. Some are remembered, but most are not because they were made unconsciously. We tend to remember the "big decisions," while seldom realizing that a single big decision is merely the sum of the many little decisions made along the way. We give just a little here and again a little there, and eventually, we have pointed ourselves in a new direction.

It is imperative that we take personal responsibility for our words and deeds because the first step toward sustainability begins with the respect and quality of care we give ourselves. Thereafter, we must extend that respect and care to each other—our families, friends, and neighbors—because a community's sustainability is measured by how well people treat one another and thus protect the global commons as everyone's birthright.

8.5.1.2 Inter-personal

Not enough can be said for civility, respect, cooperation, and hospitality toward other people. If we use these basic human behaviors to frame our decision-making, we can focus more on the mutual values that bind us and less on the tensions between our beliefs and attitudes. In practicing self-awareness and empathy, you might ask these: Would I like to be treated this way? How would I like to be on the receiving end of my own comments or actions?

If, for example, someone is rushing blindly to get somewhere and shoves you out of the way, you have at least two choices in how to respond. You might express resentment and anger (*react* in resistance), or you can be patient, kind, and understanding (*respond* in peace). Consider that you have no idea why that person was in such a hurry. What if they had just gotten an emergency phone call about a loved one in danger?

Our thoughts and actions are the seeds we sow each time we make a choice, and they will come back to us sooner or later. If we do not like the outcome of our decision, we can always choose again. In fact, while we always have a choice, we must choose—in that we have *no choice*. We are not, therefore, victims of our circumstances but rather the cumulative product of our choices and decisions. And the more we are able to choose love and peace over fear and violence, the more we gain in wisdom and the more we live in harmony and cultivate social-environmental sustainability as a regular mindset or intrinsic ethic. This is true, because what we choose to think about determines how we choose to act, and our thoughts and actions set up self-reinforcing feedback loops—or self-fulfilling prophecies, as it were—that become our individual and collective realities. Thoughts matter. As the adage goes, what you focus on will grow.

Perpetuating outmoded methods and circular thinking based on unbridled profit competition through resource exploitation inevitably lead to a dead end for people and the environment. Our addiction to competition stems from the insatiable need for perceived status; more is never enough. However, our addiction to "more" contributes to the vicious cycles of not having enough for members of non-industrialized nations, our extended family of the global community.

Competition may also involve insisting we are "right" at the expense of someone else's having to be "wrong." This insistence may lead us to vehemently defend a point of view as if another's viewpoint represents a threat to our status or our sense of material survival. There are, however, as many points of view as there are people, and each person's honest perception is indeed right from his or her vantage point. Therefore, no resolution is possible when people are committed only to winning an agreement with their respective positions. The alternative is to recognize that right versus wrong is a value judgment about human values and not a winnable argument.

At the moment it is made, each decision has a particular outcome that is more or less acceptable to a particular individual or society but not to everyone. It is best in discussion to define and focus on the principle involved as the fundamental issue. A community issue, sometimes perceived as a crisis, becomes a question to be answered. In struggling toward the answer, both positive and negative options become apparent. Once options become apparent, they become choices. Once a community has made a choice, they have agreed to a decision. This decision leads to actions, which then transform into outcomes over time.

We must, therefore, approach decision-making within the context of a sustainable landscape design by mutually supporting our common desire for a good quality of life across society. Too many efforts stagnate and wither because people fail to nurture the social nature of human interactions. Leadership that builds truly sustainable communities must include celebrations and create opportunities simply to have fun together. A community that works and plays together builds cohesion and decision-making capacity that support short- and long-term partnerships to achieve the sustainable vision of the community. Thoughtful leaders and group participants together transform into knowledgeable decision-makers who build and foster a sustainable environment and its group processes.

Strong partnerships require reciprocity, which is, after all, the essence of sustainable social networks and communities. Leaders must therefore pay particular attention in their decision-making to address the necessities and concerns of everyone in the group. People are more likely to give freely when they feel safe—listened to and respected. This kind of caring and trust and the protection of one another's dignity and humanity inspires tremendous tenacity and creativity toward resolving problems and accepting inevitable setbacks. Leadership in the form of true partnerships is a discipline that requires time and patience, as well as humility and an open, unbiased mind. We must move beyond thinking in (and applying out loud) value judgments ("good" or "bad") regarding complex ecological processes involved in nature and leave politics (which come and go) out of the discussion because it clouds the facts and our ability to be open to learning that nature's impartial neutrality is expressed through its inviolable biophysical principles.

Social-environmental sustainability can only be created on a person-to-person basis. It must grow from the bottom up and cannot be imposed from the top down. Social-environmental leadership is an ongoing, ever-evolving process of decision-making, not a fixed endpoint. The purpose of such leadership is to help people and communities empower themselves and one another as they struggle toward sustainability. Social-environmental leadership requires immature self-centeredness to be transformed into mature, other-centered teamwork. The process requires us to set aside our egos and accept diverse points of view as negotiable differences while striving for the longer-term common good. Successful and resilient teamwork supports cooperation, coordination, collaboration, possibility thinking, and issue resolution.

But even if we exercise personal discipline in designing sustainable solutions toward social-environmental problems, many of us have become far removed from the land that sustains us. Attention is focused instead on a chosen product, the success or outcome of management efforts, and anything diverted to a different result may be superficially perceived as a challenge to the economic bottom line. It is necessary, then, to reevaluate the philosophical assumptions underlying our notions of nature, decision-making, partnership, community, and society and how they can be sustainably integrated into the common future of leadership that lies ahead—across generations.

8.5.1.3 Between People and the Environment

Sustainability means that development programs must, to the extent possible, integrate the local community's requirements, desires, motivations, and identity in relation to the surrounding landscape. It also means that local people involved in development initiatives affecting the immediate environment must participate equally and fully in all debates and discussions, from the local level to the national. Here, a basic principle is that programs must be founded on local requirements and cultural values *networked* with those of the broader world, which includes understanding regional and national environmental issues and processes, long-term ecological trends, and their vision's social-environmental ramifications—all the while honoring the limitations of nature's inviolable biophysical principles and their counterpart, inviolate guidelines of decision-making. After all, social-environmental sustainability is a reciprocal relationship between people and the land. As we nurture the land, we nurture ourselves. As we abuse the land, we abuse our own well-being and threaten future generations.

Consequently, short-term trends must be viewed in relation to a much longer time frame than we may be used to conceptualizing. The more we strive to understand the potential outcomes and impacts of past, present, and future decisions on the land, our water catchments (commonly referred to as watersheds), and our communities the more informed will be our decisions.

People select a community for what it offers them within the context of its landscape. Prudent social-environmental decision-making requires an understanding that a community's location defines its character. A logging community is set within the context of the forest, a ranching community within the context of lands for grazing, and a fishing community within the context of a shore- or coastline. The specific qualities of the particular setting define the community's uniqueness. By the same token, the values, decisions, and developmental practices of a community alter the characteristics of its surrounding environment. Each community has physical, cultural, and political qualities that make it unique and more or less flexible. The degree of flexibility of these attributes in a community is important because sustainable systems must be ever-flexible, creative, and adaptable. The process of sustainable development must therefore remain flexible because what works in one community, may not work in another, or may work differently for various reasons.

Much of what a community says about itself is uniquely reflected in the aesthetics of the physical structures of its surroundings. This includes land-use, zoning, and building codes; design of transportation systems—or lack thereof; and the conservation, accessibility, and abundance of such natural assets as parks, forests, open space, streams, springs, wetlands, and rock outcroppings within the boundaries of the town or city. A community's history and evolving worldview define its collective values and how it treats the surrounding landscape. As the land is altered, the community's biophysical and social options are altered. A community and its landscape are engaged in a mutual, self-reinforcing feedback loop of reciprocity.

Local people are the power of sustainable development. As they participate in their initiatives, they grow in knowledge, self-realization, self-definition, self-determination, and decision-making capacity and skill. Such personal growth opens the community to its own evolution within the context of the people's sense of place, as opposed to coercive pressures applied from the outside.

Sustainable development encompasses any process that helps people meet their requirements, from self-worth to food on the table, simultaneously creating a more biophysically and culturally sustainable and just society for the current generation *and* those that follow. Due to its flexibility and openness, sustainable community development is perhaps more capable than other forms of development of creating such outcomes because it integrates the requirements of a local community with those of the immediate environment and surrounding landscape, as well as neighboring communities.

8.5.1.4 Between People in the Present and Those of the Future

We, the citizens of our countries and the world, must ask how our decisions and actions will affect the future. Meeting the implied obligation requires a renewed sense of identity, purpose, belonging, and consciousness—to be other-centered in caring for the welfare of those to come, as we are able to move beyond the limited knowledge and awareness of those before us.[4]

8.5.2 Guideline 2: All Relationships Are Inclusive and Productive of an Outcome

Clearly, all relationships are productive simply because they produce an outcome of some sort. The challenge for communities is what kind of outcome will result from our decisions. Because we are interconnected and interdependent, every functional part of government at every scale is influenced by how the other parts function.

Relearning—dredging the recesses of memory for things long forgotten—and the accumulation and practice of new knowledge contribute to multidimensional learning that encompasses both theoretical and practical conceptualization, decision-making, action, and the deeper aspects of reflection—including the intellect, intuition, and imagination. Multidimensional learning is important because overemphasis on action—one part of which is competition—simply reinforces fixation on short-term, quantifiable results. Our overemphasis on action can preclude reflection—a persistent practice of deeper learning that can produce wiser decisions and thus far better outcomes over time.

8.5.3 Guideline 3: The Only True Investment Is Energy from Sunlight

It must be understood and accepted by all decision-makers that the only true investment of energy in our world is light from the sun, which is 93,205,679 miles from Earth yet delivers 120 trillion watts of energy to Earth's surface—thereby providing enough energy in one hour to satisfy the requirements of the global human society for an entire year.[5] Everything else is either a *re-allocation* or *re-investment* of energy—including every transaction in the global marketplace. This makes solar energy the only truly renewable resource. Moreover, capturing this energy and converting it to usable forms is dependent on viable communities of green plants worldwide. After all, fossil fuels are at least partly ancient solar radiation archived in the Earth over millennia. It is, therefore, critical that decision-makers understand and accept that all other resources are exhaustible; they obey the three thermodynamic laws explained in Guideline 4. Until now, humanity has acted as if all resources, other than sunlight, are inexhaustible.

The elimination of one resource affects many others in the system. For example, the overexploitation of large, marine, apex predators, such as sharks and tuna, may allow the populations of smaller, plankton-feeding prey fishes to proliferate. At some point, their numbers can become large enough to dramatically reduce the number of photosynthetic phytoplankton and thus the ocean's ability to capture sunlight and absorb carbon dioxide. This ecosystem service maintains cooler waters—affecting global warming.[6] ("Phytoplankton" is from the Greek *phyton*, "plant," and *planktos*, "wanderer" or "drifter." Plankton, in turn, refers to plants and animals, generally microscopic, that float or drift in great numbers in fresh or salt water, where fish and other aquatic animals eat them.)

That said, restrictions on fishing the predators until their populations are larger than those required to maintain a sustainable yield and thus control the plankton feeders could, within biological limits, contribute to sustainable fisheries.[7] Meanwhile, warming oceans affect the major wind patterns, which affect the direction of ocean currents, which shifts oxygen-depletion (or eutrophication, popularly referred to as dead zones) in the oceans; warming water and other factors, including pollution from land, cause these zones to grow.[8] In areas depleted of dissolved oxygen mobile aquatic life must either flee or suffocate. Moreover, what happens in the oceans affects global climate, which affects the biosphere including humankind.

8.5.4 Guideline 4: All Relationships Involve a Transfer of Energy

Humans transfer energy from one place to another, from one level of intensity to another, often without understanding the laws of thermodynamics and how they affect leadership.

The *first law of thermodynamics* states that the total amount of energy in the universe is constant, although it can be transformed from one form to another. Therefore, the amount of energy remains entirely the same, even if you could go forward or backward in time. For this reason, the contemporary notion of either "energy production" or "energy consumption" is a non sequitur.

The *second law of thermodynamics* states that the amount of energy in forms available to do useful work can only diminish over time. The loss of available energy to perform certain tasks thus represents a diminishing capacity to maintain order at a certain level of manifestation (say a tree), and so increases disorder or entropy. This "disorder" ultimately represents the continuum of change and novelty—the appearance of a different, simpler configuration of order, such as the remaining ashes from the tree when it is burned.

In turn, the *law of maximum entropy production* simply means that, when any kind of constraint is removed, the flow of energy from a complex form to a simpler form speeds up to the maximum allowed by the relaxed constraint.[9] We, in contemporary society, are all familiar with the basis of this law even if we do not understand it. For example, we are probably all familiar with the fact that our body loses heat in cold weather, but our sense of heat loss increases exponentially when wind chill is factored in. This is because our clothing has ceased to be as effective as a barrier to the cold as it was before the wind chill became an issue. Moreover, the stronger and colder the wind, the faster our body loses heat—the maximum entropy of our body's energy whereby we stay warm. If the loss of body heat is not constrained, hypothermia and death can ensue.

Founded on research by scientists over the 18th and 19th centuries, the first and second laws of thermodynamics and the law of maximum entropy production meld to form the overall, unifying law of the universe—encompassing all subordinate principles, both biophysical and social. With respect to the functional melding of these three laws, Rod Swenson of the Center for the Ecological Study of Perception and Action, Department of

Psychology, University of Connecticut, says these three laws of thermodynamics "are special laws that sit above the other laws of physics as laws about laws or laws on which the other laws depend."[10]

In other words, systems are by nature dissipative structures that release energy by various means but inevitably by the quickest means possible. Suppose, for example, you live in a cold climate and have a wood-burning stove in your home with which you heat the living space. To keep your house at a certain temperature, you must control the amount of energy extracted from the wood you burn. There are nine ways to accomplish this from the type of wood you choose and so forth. The main point is that these nine (seemingly independent) factors and decisions coalesce into a synergistic suite of relationships, whereby a change in one automatically influences the other eight facets of the speed by which energy from the burning wood escapes your house. To reiterate, the first and second laws of thermodynamics and the law of maximum entropy meld to form the overall unifying law of the universe wherein all subordinate principles, both biophysical and social (decision-making), are encompassed.

8.5.5 Guideline 5: All Systems Are Based on Composition, Structure, and Function

It is imperative for a decision-maker to understand that, like every system, the organization they create is based on the composition of people and their various areas of expertise and levels of psychological maturity, which defines how the organizational system can and will function. Grasping this dynamic is critical because the degree to which organizational members are behaviorally functional or dysfunctional, broad- or narrow-minded in their points of view, team players or ego-driven grandstanders, will determine the nature of the team and the kind and quality of its products. Therefore, a leader who is conscious enough to make social-environmental sustainability the bedrock of their administration must begin with detailed attention to the roles, overall makeup, and functions of their team. This accomplishment necessitates working backward to find the interdisciplinary expertise and skill sets (composition) required for creating the necessary structure that will support the individual project of social-environmental sustainability.

8.5.6 Guideline 6: All Relationships Have One or More Trade-offs

Every decision one makes involves trade-offs. An agricultural field in which corn is grown as biofuel provides a simple example. "Corn-based ethanol is the worst among the alternatives that are available at present, although this is the biofuel that is most advanced for commercial production in the United States."[11] Corn, it turns out, is one of the most energy-intensive crops when it comes to the amount of inputs—fertilizer, water, and pesticides—it requires. The pesticides and herbicides required to produce these vast monocultures are also gas-based petrochemicals. Moreover, there is a substantial amount of diesel fuel needed to operate the farm machinery, which says nothing of the enormous quantity of water this exceedingly thirsty crop requires—1,700 gallons for each gallon of ethanol produced.[12] Therefore, increased reliance on biofuel is merely trading an oil problem for a water problem that goes from the field into the cities and into the oceans of the world.[13]

On a global scale, the alterations in surface-water chemistry from the human deposition of nitrogen, sulfur, and dissolved inorganic carbon are relatively slight compared with the acidification caused by the oceanic uptake of anthropogenic carbon dioxide. The impacts are more substantial in coastal waters, however, than in the deep ocean. In coastal areas, the ecosystem's response to acidification has severe implications for people, especially those who rely on shellfish harvest.

8.5.7 Guideline 7: All Systems Have Cumulative Effects, Lag Periods, and Thresholds

A decision-maker must help the public to understand that all systems have incremental, *cumulative effects* that occur below our level of awareness (this is known as *hysteresis*, which refers to the time lag between a cause and an effect) until such a time that sufficient change has taken place to cross a *threshold of visiblity*. In a human sense, this often means accepting the delayed gratification of a vision's outcome, despite wanting to see immediate results. Herein lies a major challenge for leaders of industrialized countries, where the citizenry insists on and uses technology that constantly ramps up the speed of transactions. The faster technological systems work, the faster people want them to work.

Unfortunately, we in the Western industrialized countries—and the United States in particular—are too often like the rich but foolish man in the Buddha's parable who wanted instant results:

> Once there was a wealthy but foolish man. When he saw the beautiful three-story house of another man, he envied it and made up his mind to have one built just like it, thinking himself just as wealthy. He called a carpenter and ordered him to build it. The carpenter consented and immediately began to construct the foundation, the first story, the second story, and then the third story. The wealthy man noticed this with irritation and said: "I don't want a foundation or a first story of a second story; I just want the beautiful third story. Build it quickly."[14]

Everything in nature has its own timing. We would be wise to accept that timing with patience, including the patience of achieving consensus prior to committing an action that will affect all generations to come. Many people in the industrialized nations seem to have the attitude that "time is money" and so are in a hurry to act, often without thinking through the potential consequences of their actions on either people or the environment. Therefore, while we have control of what we introduce into the environment, once introduced, the changes may likely be forever beyond our control. This realization must be accepted by every decision-maker and leader. To our greatest ability, we must examine options, analyze the options for benefits and risks not just to ourselves but to the system and its heirs, monitor the decision as applied over time, and be willing to respond. Finally, we must be willing collectively to accept responsibility for the outcomes of our decisions, including unforeseen consequences.

Some years ago, while working with the same Shinto priests in Japan, which I (Chris) mentioned earlier, I learned a valuable lesson. The priests spent much time discussing the pros and cons of various decisions, as well as the potential outcomes of this choice or that. While the meetings seemed interminable and inefficient to me, when the priests came to a consensus, they acted with their collective wisdom because they had winnowed the possibilities and probabilities through an informal but very effective risk analysis. This experience taught me that taking as long as the process requires to do well, with patience and keen attention, is often far more valuable than instant gratification because doing it wisely the first time is both effective *and* efficient since it precludes the necessity of having to do it over at the extra cost of time, labor, and capital (social and financial). Thus, we have a far greater probability of achieving our desired outcome both efficiently *and* effectively, by giving the process its due the first time. But first, we have to properly understand the notion of *leisure*.

Whereas we tend to think of leisure as the privilege of the well-to-do, Brother David Steindl-Rast (a Benedictine monk) reminds us that leisure

> is a virtue, not a luxury. Leisure is the virtue of those who take their time in order to give to each task as much time as it deserves. . . . Giving and taking, play and work, meaning and purpose are perfectly balanced in leisure. We learn to live fully in the measure in which we learn to live leisurely.[15]

Therefore, decision-makers must learn and teach leisure and patience by example if their legacy is going to be a true benefit for all generations. This may be one of the most challenging guidelines of all when communities are faced with a regulatory or policy deadline meted out from their governance structure, which is frequently the case with such decisions as land-use zoning. There must be allowances for the necessary time for community members to come to a reasoned, well-informed decision and consultations (including with experts).

8.5.8 Guideline 8: Change Is an Irreversible Process of Eternal Becoming

One of the more complex issues of decision-making—especially social-environmental decisions—is dealing with the fixed nature of change as a constant process. All systems are cyclical, but none are perfect circles, which often confronts decision-makers with perplexing choices because sometimes people seem to want events to go perpetually their way. For example, industrialized nations and their citizenry insist that the economic system be linear and ever-growing (one vision of capitalism). This unrealistic vision has produced a symptomatic configuration based so completely on competition and boundless growth that it is destroying the global environment's ability to serve humanity as a viable life-support system. The influence of this three-century speculative legacy on public policy and corporate decision-making has run its course; now we need to move in a less selfish and more inclusive direction.

It is imperative that people become aware of the long-term effects of their decisions. We say this because children are one of the two silent parties in all social-environmental decisions; the land, with its systems and its resilience, is the other. Leaders, at whatever level of society, must understand the social, environmental, and economic circumstances to which they are committing future generations through their decisions of today. If the outcome of their decisions and actions is a deficit in terms of the future in terms of the productive capacity of ecosystems to serve human necessities, it is analogous to taxation without representation.

What today's and tomorrow's decision-makers must increasingly understand is that human-induced climate change depends not only on the magnitude of the change but also on the potential reversal of the process of change, which will take about 1,000 years after the emissions cease due to the slower cooling of the world's expanding and warming oceans.[16] Even with the potential for a reversal of global warming as a process, the current and ongoing biophysical consequences of climate change are already irreversible.

8.5.9 Guideline 9: Systemic Change Is Based on Self-Organized Criticality

According to a theory called self-organized criticality, the mechanism that leads to a minor event (analogous to the drop of a pin) is the same mechanism that leads to a major event (analogous to an earthquake).[17] Not understanding this, analysts have typically and erroneously blamed some rare set of circumstances (some exception to the rule) or some powerful combination of mechanisms when catastrophe strikes.

Self-organized criticality is an important concept for decision-makers to understand because most current leaders base their decision-making on linear, symptomatic thinking. They do not realize that virtually all social collapses are initiated from within, even if triggered from without (how could it possibly be otherwise unless each nation lived in a bubble, which it does not?). Consider that all civilizations—past and present—evolved by similar steps: growth of knowledge and technology through discoveries and inventions and through the ideas of law and government, family, and property, all of which are based on observation and the accumulation of experimental knowledge. As such, all civilizations have much in common, and their evolutionary stages are connected in somewhat similar sequences of cultural development.

The arts of subsistence and the achievements of technology can be used to distinguish the periods of human progress. People began their journey by gathering fruits, nuts, and seeds. Then they learned to hunt, fish, and use fire; invented the spear and atlatl, advancing to the bow and arrow, and then gunpowder. They developed the art of making pottery, learned to domesticate animals and cultivate plants, began using adobe and stone in building houses, and learned to smelt bronze and iron and use it in making tools. Finally, "civilization" began with agriculture[18] and the invention of the phonetic alphabet, culminating in all today's wonders of human accomplishment.

Earlier civilizations have been marked by birth, maturation, and demise, the last commonly caused by uncontrolled population growth that outstripped its sources of available energy, especially during times of climate stress, which initiated internal and external conflicts. In some cases, survivors moved to less-populated, more fertile areas as their civilizations collapsed. Today, however, there is nowhere left on Earth to go.

Yet having learned little or nothing from history, the rich, industrialized nations are engaged in the competitive race of destruction of the very environments from which their societies sprang and on which they—and all inhabitants of the biosphere—rely for continuance. The separation of work from the rest of life, which began many centuries ago with the inception of agriculture, is today fully manifested. Jobs are so fundamental to people's sense of identity, self-sufficiency, and social position that a job loss can cause depression and stress and long-term stress damages health.[19]

In contrast, cooperation and coordination with sharing and caring reduce the perceived need to compete for survival and social status, except in play—and perhaps storytelling. Linking individual well-being strictly to individual production and competition exacerbates social inequality, poverty, and the environmental degradation inherent in unbridled materialism. Self-centered acquisitiveness is not an inherent trait of our species but is acquired based on social pressure and expectations leading to the insecurity that fosters our sense of personal power: that we matter and we can make a positive difference. Therefore, industrial society's present way of life cannot be the final evolutionary stage for humanity. What is the next frontier for civilization? Is it outer space as so often implied? No, it is not outer space. It is inner space—mastery and maturity of the self, which is life's most important calling. As the Buddha said, "Though he should conquer a thousand men in the battlefield a thousand times, yet he, indeed, who would conquer himself is the noblest victor."[20]

Sociologists, on the other hand, might think of self-organized criticality as the butterfly effect, so called because it is analogous to the flapping wings of a butterfly, which represents a small change in the initial condition of the system that in turn causes a chain of occurrences leading to large-scale alteration of circumstances. Framed in the positive, this trend means that what a few influential, psychologically mature decision-makers think and say can spread and bring about significant changes in the thinking and behavior of large numbers of people, which in turn would affect the social-environmental sustainability and inheritance of all generations.[21] As the adage (frequently attributed to anthropologist Margaret Mead [1901–1978]) goes, "Never doubt that a small group of thoughtful, committed citizens can change the world; indeed, it's the only thing that ever has."[22]

8.5.10 Guideline 10: Dynamic Disequilibrium Rules All Systems

All social systems, like those of nature, are ruled by dynamic disequilibrium, which simply means that the universally constant process of change, with its ever-novel outcomes, precudes any fixed state of lasting balance. Therefore, the notion of any kind of steady-state economics, as it is known on the Internet, is impossible. Here, the abiding paradox of life is that we have a choice in everything we think and almost everything we do—except practicing relationships, experiencing ourselves as we experience relationships, choosing, changing the world, living without killing, and dying. In those, we have no choice of what we do, but we do have a choice of how we do it—and we must choose. Moreover, we must understand that not choosing is still a choice. In addition, we make a new choice (even if it is doing nothing) each time we experience a change in life circumstance, which, of course, is an ongoing process, be it the outworking of biophysical principles that govern life or how we view the life changes as we mature in years.

The constancy of change dictates the omnipresence of choice. Life can therefore be viewed as an eternal parade of decisions, each of which marks a fork in the path. Each time we decide, we forgo other choices—the trade-offs mentioned earlier. Nevertheless, each decision creates a kaleidoscope of additional choices. In turn, choice leads to both wisdom and folly manifested in the ensuing consequences. In a 1970 speech in London, Israeli statesman Abba Eban (1915–2002) said, "History teaches us that men and nations behave wisely once they have exhausted all other alternatives."[23] The 21st century is the time to let go of our resistance to change and find the humility to learn from nature by accepting and living within the inviolable constraints of nature's biophysical principles.

In this sense, everything we think and do has a trade-off of positive *and* negative consequences at the time we form the thought, decide, and act. Hence, each choice—and the subsequent decision to act—involves the trade-offs of hoped-for outcomes amid the irreversibility of life's unknowns and uncertainties that are the essence of life's *dynamic disequilibrium*.

8.5.11 Guideline 11: Success or Failure Lies in the Interpretation of an Event

All relationships are self-reinforcing feedback loops that, are neutral in valuation in nature, where all values are intrinsic. However, these same feedback loops are positive, negative, and frequently mixed in terms of human valuation because we seek specific, predetermined outcomes to provide the illusion of being in control of our circumstances. This human dynamic is the same as that driving the notion of success or failure, which lies in the interpretation of an event—but not the event itself, something every public decision-maker would do well to learn.

To illustrate, I (Chris) was once asked to mediate a conflict in Northern California over how to "restore" the Mattole River. During the process, some of the long-time, older residents began lamenting how newcomers into "their" valley had destroyed "their" river through years of overuse and abuse. Finally, a youth in his late teens, who had been in juvenile detention, spoke up and said, in effect, "I don't know what you're talking about. I've been working for three years with a crew to improve the river's condition. The river is so much better than when I started. What's your problem?"

The only thing the old-timers could see was the loss of a resource they esteemed to be in "pristine condition," whereas the boy perceived a vast improvement in a short period of time. The success or failure of the efforts to heal the river was therefore perceived differently, depending on the time frame of life experiences and direct experience of those involved. In this case, the old-timers constantly reinforced their collective grief over the "negative" changes for which they blamed others, although they had been complicit in rendering the changes through their own actions. The boy, on the other hand, was encouraged by his ability to produce a positive change in the river's condition—to which the old-timers were blind but for which he gladly took responsibility. In this case, to the old-timers, the glass was half empty (failure), whereas the boy saw the same glass as half full (success), while the water in the proverbial glass was unchanged.

8.5.12 Guideline 12: People Must Be Equally Informed If They Are to Function as a Truly Democratic Society

For a group of people to be socially functional, they must be given the opportunity to be equally informed about what is going on and how it is projected to affect them. In other words, there must be no secrets or hidden agendas that potentially exclude or can be detrimental to any member. Inequality of any kind within the group related to gender and social class is merely the fear of inadequacy disguised as a privilege by those who would impose their will. This particular guideline poses vast implications beyond the obvious. Implications include affordable access to technology; communication; high-quality, high-speed internet services regardless of whether citizens are in urban, rural, or remote settings—particularly online education from grade school through university; medicine and health care; weather, fire, flood, or other emergency preparedness; cultural amenities; and other information concerning food and agriculture, housing, and so forth. An increasingly important implication concerns personal empowerment through informed decision-making in the form of voting in elections. This theme has been explored in various books and research articles and is often referred to as the Network Society.[24]

8.5.13 Guideline 13: We Must Consciously Limit Our "Wants"

We who reside in industrialized nations find ourselves ensnared in the process of selling and buying things in the marketplace to the extent we cannot readily imagine human life being otherwise. Even our notion of well-being and of despair are wedded to the markets' flow and ebb. Why is this so much a part of our lives? It is largely because people have yet to understand the notion of conscious simplicity, which is based on the realization that there are two ways to wealth: want less or work more.[25] Put differently, wealth lies in the scarcity of one's wants as opposed to the abundance of one's possessions. I (Chris) learned the truth of this while in the desert of Egypt in the 1960s. I poured a little water from my canteen down a cistern well the Romans had built and watch it disappear into the sand. I instantly realized, as it disappeared, that a well cannot be filled from the top by pouring water in, but only from the bottom if the belowground water table is high enough. The analogy is that no human can fill a sense of inner emptiness by garnering possessions from without—only by finding spiritual fulfillment from within.

By consciously examining and limiting our wants, it is likely that we can have enough to provide comfortably for our necessities as well as some of our most ardent desires—and leave more for other people to do the same. Unfortunately, capitalistic systems tend to contribute to dissatisfaction and a continual stimulus to purchase superfluous items at the risk of personal debt, the long-term expense of the environment, and thus generational impoverishment. Herein lies one of the greatest challenges of contemporary decision-makers, a challenge that must be met if our society—and the environment that supports it—is to survive the 21st century with any kind of dignity and well-being for more than the wealthy top 1%. As of late 2021, in the United States this group of 1.3 million households, earning in excess of

$500,000 annually, holding more income than the entire 60% middle-income households.[26]

8.5.14 Guideline 14: Simplicity Is the Key to Contentment, Adaptability, and Survival

Simplicity in living and dying depends on appreciating things as simple, small, sublime, and sustainable. What is more, simplicity is the key to contentment, adaptability, and survival as a culture. Beyond some point, complexity becomes a decided disadvantage with respect to cultural longevity, just as it is to the evolutionary longevity of a species. As painter Hans Hoffman (1880–1966) stated, "The ability to simplify means to eliminate the unnecessary so that the necessary may speak."[27] To this, His Holiness the 17th Gyalwang Karmapa Ogyen Trinley Dorje would add,

> A life full of material goods and barren of compassion is quite unsustainable from an ecological and karmic point of view. Of course, advertisements are always telling us that the path to happiness lies in purchasing the goods they sell. . . . We don't have to live a life that is sold to us—we can make the brave choice to live simply.[28]

The cattle ranchers I (Chris) used to work for as a young man were of the opinion that "[i]f you have your health, you have everything." That notion goes a long way in defining the simplicity of true wealth.[29]

8.5.15 Guideline 15: Nature, Environmental/Cultural Wisdom, and Human Well-Being Are Paramount

To summarize, placing material riches, as symbolized by the money chase, above the wealth of nature, spirituality, and human well-being is the road to social impoverishment, environmental degradation, and the collapse of societies and their life-support systems—as demonstrated throughout human history. We must, therefore, rethink our priorities and place the viability of ecosystems at least on par with the economics of "natural capital," which means reconnecting with nature, if our own well-being is to prevail.

Here, it is instructive to consider the difference between money and the true wealth of nature's services and our personal well-being. Conventional money knows no loyalty to a sense of place, person, local community, landscape, region, or even a nation, so it flows toward a global economy in which traditional, social bonds give way to a rootless quest for the highest monetary return at virtually any social-environmental cost. The deeply profound price we pay for money is the hold it has on our (and the next generations') sense of what is possible—the prison it builds around our imaginations. This sentiment is reminiscent of an aphorism often attributed to journalist Sydney J.

Harris (1917–1986): "Men make counterfeit money; [but] in many more cases, money makes counterfeit men."[30]

One of the most important indicators of economic stability is social-environmental sustainability, which means not only quality interpersonal relationships but also quality reciprocal relationships between people and their environment. A truly viable economy is based on love and reciprocity, where people do kind and useful things for one another with no expectation of financial gain. Such mutual caring is the soft social capital that both creates and maintains the fabric of trust, which in turn is the glue of functional families, communities, and societies.

According to Bernard Lietaer, of the Center for Sustainable Resources at the University of California at Berkeley,

> Money is like an iron ring we've put through our noses. We've forgotten that we designed it, and it's now leading us around. I think it's time to figure out where we want to go—in my opinion toward sustainability—and then design a money system to get us there.[31]

Pathology and its subsequent dysfunction enter the economic system when money, derived as a convenient means of exchange, becomes the factor that defines the purpose of life for individuals and their communities. As the soft social capital of mutual caring dwindles and the resulting quality of family life withers, a community's members typically become increasingly apathetic or competitive. In turn, as human relationships become more and more dysfunctional, a community's infrastructure crumbles into ever-greater disarray at an ever-increasing social-environmental cost to all generations.

In regard to sustainable economic systems, however, a decades-long global discussion continues to evolve across generations and is led by scientists, authors, communities, international organizations (such as the United Nations), and individual countries themselves.[32]

8.5.16 Guideline 16: Every Legal Citizen Deserves the Right to Vote

Every legal citizen of every country deserves the right to an equal vote of their conscience on how their country is to be governed because they and their children and their children's children must live with the consequences of the collective choices and actions.

8.5.17 Guideline 17: This Present Moment Is All We Have

This eternal, present moment is all we ever have in which to act. The past is a memory, and the future never comes. *Now* is the eternal moment. The inviolable guidelines of decision-making are enshrined in every moment of everyday life, whether we recognize them or not. As such, the leader who honors them will move forward unafraid and, in so doing, be the keeper of everyone else's dignity by keeping their own dignity in the eternal, present moment—*leading by conscious example.*

8.6 Connecting the Guidelines on Leadership and Decision-Making

Change often occurs on the brink of disaster between need and fear. On the one hand, we know we need to do things differently. On the other hand, we are terrified of facing the unknown, unfamiliar, and uncertain. To change our direction for the future, however, we must suspend our conventional notions about change and our ability to learn because there are no problems to resolve other than those we perceive as manifestations of how we think and act. The problems we face are a matter of who we are consciously. Many people prefer to err again and again rather than let go of some cherished belief, pet notion, deified assumption, or staunchly defended position. Others err because they are pessimistic in their outlook and are thus blind to viable options.

Social-environmental sustainability demands that decision-makers go beyond the immediate human valuation of a given resource to examine and disclose the fundamental issue of how its use will affect the long-term, biophysical sustainability of the ecosystem of which it is a component. One must also recognize and disclose the long-term, social-environmental issues that need to be dealt with concerning the method by which a resource is extracted. This is necessary because the overall integrity of an ecosystem, its productive capacity, and the sustainability of its resources will determine the array of options passed forward to future generations.

But when we refuse to accept a lesson from nature's biophysical principles, it forms an intellectual blockage in our potential understanding of those principles that ultimately govern social-environmental sustainability. The inertia of informed denial is the usual procedure when the immediate economic cost of rectifying a mistake is thought to be great, which is nothing less than passing the debt to some other generation. Informed denial, as a remedy, is based on the same level of consciousness and its decision-making that initiated the problem in the first place (a reaction instead of a response), which can only compound the problem.

The refusal to accept the biophysical constraints of nature as they affect today's community-oriented decisions is a serious issue in Western industrialized countries and those adopting Western ways, particularly when it comes to the control of how energy—other than that of solar radiation—is used.

To accommodate the sustainability of the environment's ecological integrity, each person who would lead must understand the leadership guidelines and the biophysical principles on which they are based. These principles form the underpinnings through which nature operates and the social limitation we must understand and accept if we are to participate with nature in a sustainable fashion. Moreover, social-environmental justice dictates that leaders be held accountable for the outcome of their decisions (preferably while in their position of authority instead of after the fact) because the effects of their decisions become the consequences for all generations—those present and those yet to come. This is particularly poignant in the face of an exploding human population, rapidly degrading ecosystems, and thus dwindling per-capita resources.

8.7 References

1. Joel E. Correia. 2022. Why the UN's Recent Declaration on Human Rights Should Be Taken Seriously. *The Conversation* (August 5, 2022). www.fastcompany.com/90776402/why-the-uns-recent-declaration-on-human-rights-should-be-taken-seriously (accessed September 6, 2022).
2. National Geographic Resource Library. *Citizen Science*. https://education.nationalgeographic.org/resource/citizen-science (accessed January 21, 2023).
3. Euroscientist. *Science Together—Contributing to Citizen Science Projects Around the World*. www.euroscientist.com/citizen-science-around-world/ (accessed January 21, 2023).
4. Chris Maser. 1992. Do we Owe Anything to the Future? In: *Multiple Use and Sustained Yield: Changing Philosophies for Federal Land Management?* Proceedings and summary of a workshop convened on March 5 and 6, 1992, Washington, DC., pp. 195–213. Congressional Research Service, Library of Congress. Committee Print No. 11. US Government Printing Office, Washington, DC.
5. Charles Petit. 2009. Cold Panacea. *Science News*, 175:20–23.
6. **(1)** Georgi M. Daskalov, Alexander N. Grishin, Sergei Rodionov, and Vesselina Mihneva. 2007. Trophic Cascades Triggered by Overfishing Reveal Possible Mechanisms of Ecosystem Regime Shifts. *Proceedings of the National Academy of Sciences*, 104:10518–10523; **(2)** Peter Ward and Ransom A. Myers. 2005. Shifts in Open-Ocean Fish Communities Coinciding with the Commencement of Commercial Fishing. *Ecology*, 86:835–847.
7. **(1)** Chris L. J. Frid, S. Hansson, S. A. Rijnsdorp, and S. A. Steingrimsson. 1999. Changing Levels of Predation on Benthos as a Result of Exploitation of Fish Populations. *Ambio*, 28:578–582; **(2)** R. Q. Grafton, T. Kompas, and R. W. Hilborn. 2007. Economics of Overexploitation Revisited. *Science*, 318:1601.
8. **(1)** D. Pauly and V. Christensen. 1995. Primary Production Required to Sustain Global Fisheries. *Nature*, 374:255–257; **(2)** John A. Barth, Bruce A. Menge, Jane Lubchenco, and others. 2007. Delayed Upwelling Alters Nearshore Coastal Ocean Ecosystems in the Northern California Current. *Proceedings of the*

National Academy of Sciences, 104:3719–3724; **(3)** Ryan R. Rykaczewski and David M. Checkley Jr. 2008. Influence of Ocean Winds on the Pelagic Ecosystem in Upwelling Regions. *Proceedings of the National Academy of Sciences*, 105:1965–1970.

9. **(1)** Rod Swenson. 1989. *Emergent Evolution and the Global Attractor: The Evolutionary Epistemology of Entropy Production Maximization*. Proceedings of the 33rd Annual Meeting of The International Society for the Systems Sciences, P. Leddington (ed)., 33(3):46–53, **(2)** Rod Swenson. 1991. Order, Evolution, and Natural Law: Fundamental Relations in Complex System Theory. In: C. Negoita (ed.), *Cybernetics and Applied Systems*. Marcel Dekker Inc., New York, pp. 125–148 **(3)** Rod Swenson and Michael T. Turvey. 1991. Thermodynamic Reasons for Perception-Action Cycles. *Ecological Psychology*, 3:317–348.
10. Rod Swenson. 2000. Spontaneous Order, Autocatakinetic Closure, and the Development of Space-Time. *Annals New York Academy of Sciences*, 901:311–319.
11. Martha J. Groom, Elizabeth M. Gray, and Patricia A. Townsend. 2008. Biofuels and Biodiversity: Principles for Creating Better Policies for Biofuel Production. *Conservation Biology*, 22:602–609.
12. **(1)** Tad W. Patzek. 2004. Thermodynamics of the Corn-Ethanol Biofuel Cycle. *Critical Reviews in Plant Science*, 23:519–567; **(2)** David Pimentel and Tad W. Patzek. 2005. Ethanol Production Using Corn, Switchgrass, and Wood; Biodiesel Production Using Soybean and Sunflower. *Natural Resources Research*, 14(1):65–76; **(3)** Jason Hill, Erik Nelson, David Tilman, and others. 2006. Environmental, Economic, and Energetic Costs and Benefits of Biodiesel and Ethanol Biofuels. *Proceedings of the National Academy of Sciences*, 103:11206–11210; **(4)** Manfred Kroger. 2007. Forum: Corn Is Food, Not Fuel. *Pittsburgh Post-Gazette* (April 8, 2007); **(5)** Alice Friedemann. *Peak Soil*. http://culturechange.org/cms/index.php?option=com_content&task=view&id=107&Itemid=1 (accessed February 20, 2022); **(6)** Lian Pin Koh. 2007. Potential Habitat and Biodiversity Losses from Intensified Biodiesel Feedstock Production. *Conservation Biology*, 21:1373–1375; and **(7)** Sid Perkins. 2007. Groundwater Use Adds CO_2 to the Air. *Science News*, 172:301.
13. Robert F. Service. 2009. Another Biofuels Drawback: The Demand for Irrigation. *Science*, 326:516–517.
14. *The Teachings of Buddha*. 1985. Bukkyo Dendo Kyokai, Tokyo, Japan.
15. Brother David Steindl-Rast. 1984. *Gratefulness and The Heart of Prayer: An Approach to Life in Fullness*. Paulist Press, Ransey, NJ., Page 74.
16. Susan Solomon, Gian-Kasper Plattner, Reto Knutti, and Pierre Friedlingstein. 2009. Irreversible Climate Change Due to Carbon Dioxide Emissions. *Proceedings of the National Academy of Sciences*, 106:1704–1709.
17. Per Bak and Kan Chen. 1991. Self-Organizing Criticality. *Scientific American*, January:46–53.
18. **(1)** Stacey Y. Abrams. 1991. The Land Between Two Rivers: The Astronomy of Ancient Mesopotamia. *The Electronic Journal of the Astronomical Society of the Atlantic*, 3(2). Georgia State University Press. No page numbers given and **(2)** Stacey Y. Abrams. The Fertile Crescent. *The Electronic Journal of the Astronomical Society of the Atlantic*, 3 (2). http://visav.phys.uvic.ca/~babul/AstroCourses/P303/mesopotamia.html (accessed February 21, 2022).
19. Radha Chitale. 2009. Job Loss Can Make You Sick. *ABC News*. http://abcnews.go.com/Business/WellnessNews/story?id=7530730&page=1 (accessed June 1, 2009).
20. The Teachings of Buddha. Op. cit.

21. Jamie L. Vernon. Understanding the Butterfly Effect. *American Scientist*, 105(3):130.
22. This quote is attributed to Margaret Mead via the epigraph to Chapter 6 of Donald Keys' 1982 book *Earth at Omega: Passage to Planetization*, The Branden Press (January 1, 1982), ISBN-10: 0828317436; ISBN-13: 978–0828317436.
23. Abba Eban. Speech in London, December 16, 1970, as quoted in *The Times* (December 17, 1970).
24. (1) Kenneth L. Hacker and Jan van Dijk. 2000. *Digital Democracy: Issues of Theory and Practice*. Sage Publications. https://sk.sagepub.com/books/digital-democracy (accessed July 14, 2022) and (2) Jan van Dijk and Kenneth L. Hacker (editors). 2018. *Internet and Democracy in the Network Society*. Routledge Studies in Global Information, Politics and Society., 213 pp.
25. Ernst Friedrich Schumacher. 1973. *Small Is Beautiful: A Study of Economics As If People Mattered*. Blond & Briggs, London, 288 pp.
26. Alexandre Tanzi and Mike Dorning. 2021. After Fifty Years of Decline (see www.pewtrusts.org) America's Middle Class Now Holds a Smaller Share of U.S. wealth than the top 1%. www.bloomberg.com/news/articles/2021-10-08/top-1-earners-hold-more-wealth-than-the-u-s-middle-class (accessed July 15, 2022).
27. Hans Hoffman. 1967. *Search for the Real and Other Essays* (Rev. ed.). The MIT Press, Cambridge, MA.
28. H. H. [His Highness] 17th Gyalwang Karmapa Ogyen Trinley Dorje. 2011. Walking the Path of Environmental Buddhism through Compassion and Emptiness. *Conservation Biology*, 25:1094–1097.
29. Joshua N. Hook, Adam S. Hodge, Hansong Zhang, and others. 2021. Minimalism, Voluntary Simplicity, and Well-Being: A Systematic Review of the Empirical Literature. *The Journal of Positive Psychology*. https://doi.org/10.1080/17439760.2021.1991450 (accessed July 15, 2022).
30. Sydney J. Harris (1917–1986) was a well-known Chicago journalist, born in London, whose aphorisms became famous; however, the quote cannot be authoritatively traced to his writings.
31. Sarah van Gelder. 1997. An Interview with Bernard Lietaer. *YES! A Journal of Positive Futures*, Spring:10–12.
32. Edward B. Barbier. 1987. The Concept of Sustainable Economic Development. *Environmental Conservation*, 14(2):101–110.

8.8 Discussion Questions

1. How would you describe the characteristics of an intelligent decision?
2. Of the decision-making guidelines, which three stand out as the most important? Why?
3. Why is flexibility in thinking and planning so important to the outcome of a community's sustainable vision?

4. What does inclusive brainstorming mean to you? How important do you think it is and why?
5. Recall the passage "People who succeed in changing their lives and growing beyond their present limitations of consciousness do so because they not only know that success is the interpretation of an event and not the event itself but also because they recognize their own inner patience with themselves and thus self-control, which gives them a sense of personal authority."
6. How might a person develop and practice patience? Do you feel that you have "inner patience" with yourself? If so, what does that mean to you and how does it make you feel about yourself?

9

Time, Change, and Resilience: The Theory and Practice of Community Sustainability

9.1 Introduction

Previous chapters presented both theories and direct experience gleaned from 40 years of community process, leadership, and practice. Chapter 9 presents real-world examples, gathers some of the specific characteristics or features that are common to successful sustainable community projects, and helps us begin conversations worldwide about how we can learn from other communities' work, advance new principles and methods, and make our own projects worthy of sharing and adaptation in other locations. The chapter concludes with a summary of considerations to keep in mind for the future, then transitions to the book's conclusion.

9.2 Case Studies of Community Sustainability

The purpose of this section is to feature short descriptions of recent or ongoing sustainable community case studies from diverse locations. Sustainable community planning is in a period of historic growth and development, wherein citizens around the world are engaging local governments and businesses, scientists, planners, designers, architects, landscape architects, hydrologists, students and educational institutions, transportation engineers, public health experts, and others in the dynamic, revolutionary envisioning of sustainable possibilities and project implementation on the ground.

As communities go, not only are cities home to more than 57% of people on Earth (eight billion and counting), but cities produce more than 70% of greenhouse gas (GHG) emissions—making cities a logical driver for innovative sustainable initiatives. The present is a time of rapid change in terms of goals, rationales, and technical advances. Two challenges became obvious in identifying, reviewing, and choosing a selection of case studies for this

DOI: 10.1201/9781003353744-9

book: (1) The flux and rapid trajectory of concepts, practices, and lessons coupled with a lack of relatively centralized and accessible stories from various locations. (2) As is often the case, those deeply engaged in creating change may not have the time or the inclination to write up their groups' process and results into a published record to share globally. It may also be the case that we frequently miss the great extent to which the local *is* the global and may think our community's project and thinking are only of interest and use to ourselves and our own region. The sources of many of the case studies, therefore, are derived from public agencies, nonprofit organizations, and peer-reviewed research articles; we are grateful to the teams who produced these studies. Much, but not all, of the literature drawn upon is open access via the internet and a public library. Several more case studies are emerging, forming a broader foundation of experience from which students, citizens, and planners might identify and evaluate common elements of successful projects.

The well-known United Nations program on sustainable cities is generating and inspiring a tremendous amount of work to effectively tie advancements in urban design to the United Nations Sustainable Development Goals (UN SDGs), specifically Goal 11, which pertains to cities. For instance, the International Union for the Conservation of Nature's (IUCN) eight standards for designing, quantifying, and implementing "nature-based solutions" are also impacting contemporary community projects. (For more information on both the UN SDG 11 and the IUCN's standards for nature-based solutions, see the section "Selected Further Resources for the Reader" following the book's conclusion.) Both of the reference sites provide students and practitioners of sustainable community planning with real-world case studies.

In a nutshell, the IUCN's eight standards or criteria for the nature-based solutions are as follows:

Criterion 1: The solutions effectively address societal challenges.

Criterion 2: The solutions' design is informed by scale.

Criterion 3: The result is a net gain to the integrity of biodiversity and ecosystem.

Criterion 4: The solutions' results are economically viable.

Criterion 5: The solutions are based on inclusive, transparent, and empowering governance processes.

Criterion 6: The solutions equitably balance trade-offs between the achievement of their primary goal(s) and the continued provision of multiple benefits.

Criterion 7: The solutions are managed adaptively, based on evidence.

Criterion 8: The solutions are sustainable and mainstreamed within an appropriate jurisdictional context.[1]

In fact, a 2022 research study by Blanco et al. sought to identify convergences in sustainability themes among case study locations that attempted to incorporate or meet both the UN SDG 11 and the standards for IUCN-based solutions. The authors focused on six cases for neighborhood-scale projects that emphasized site protection from population pressures, as well as cases that seek to regenerate one or more features or functions of the local or pre-existing ecosystem. The cases are located in Oregon, United States; Mumbai-Pune region, India; Edmonton, Canada; Castlemaine, Australia; and Bretigny-sur-Orge, France; and Dijon, France.[2]

The primary value of these six projects includes the restoration of habitat and biodiversity, as well as the conservation of water, energy, and other natural resources. The projects were based on the intense evaluation of each site, to establish the project baseline, including metrics on soil, vegetation, water, wildlife, and insects. Such diagnostics allow the reclamation and restoration of historic uses (industrial sites) and biomimicry, a principle in urban design that seeks to identify and apply innovative, sustainable solutions based on nature in contrast to the former ethos within sustainability based on mitigating negatives.[3]

The strongest convergence of themes (or sustainable strategies) detected by the study authors (Blanco et al. 2022) among the six cases are as follows:

1. On-site generation of renewable energy; on-site management of stormwater
2. Reduced consumption of energy
3. Design for active living and low-carbon mobility
4. Inclusion of minor public green spaces[4]

The findings from the study by Blanco et al. (2022) includes two key points that echo the themes of this book: that the "pillars of regenerative urban projects" require diagnostic, baseline data and citizen participation. First, it is critical to undergo a diagnostics process on both the social sphere and the site involved to derive baseline data needed to outline and inform project priorities and to devise a place-appropriate list of relevant solutions and strategies to reconnect functions within the system. Morever, noting a lack of participation by area-dwellers across the six case studies, the authors emphasized the critical need for active involvement by citizens, beginning with their being educated to increase awareness about the site, its biophysical structure, and realistic, potential options arising through the (participatory) design process. They reiterated,

> Users' role in sustainable and regenerative projects is more than necessary for their final ecological performance. More than designed to be sustainable, urban projects have the challenge to help people to adopt sustainable lifestyles, accelerating societal change.[5]

This quote resonates with Jacques Yves Cousteau's observation that we protect what we love, we love what we understand, and we understand what we're taught. While understanding may not always accompany or precede love, understanding the ecosystems that surround and support us surely inspires a lifetime of heightened awe, appreciation, fascination, and gratitude—love's satellite emotions.

Cities have become leaders in sustainable development. One example is Chicago, Illinois, United States, which is home to several sustainable neighborhood initiatives, including the recent Garfield Green project. Winning the 2020 Reinventing Cities[6] design global award, Garfield Green was also one of only 11 programs selected by the City of Chicago to receive highly competed-for low-income housing tax credits. The Reinventing Cities 2020 competition called for proposals to transform unused or abandoned urban spaces, such as buildings, parking lots, and vacant spaces into innovative spaces to counteract climate change. The Garfield Green project began on a vacant 1.5-acre lot. It is implementing modular construction of affordable housing and public space, including a plaza. Garfield Green's planned features include a green roof that will grow food and capture stormwater runoff, energy derived from solar panels, accessible public transit, and public art. Interestingly, when first identified by the City of Chicago as a potential applicant for the Reinventing Cities 2020 round, there was little citizen involvement. However, two years later, several different citizen committees have become engaged and involved under the initiative of a development organization Preservation of Affordable Housing.[7] Garfield Green is a long-term, ambitious project from which we will be able to learn much as we watch it unfold.

Curitiba, Brazil, South America, is a city of two million people with an impressively long history of sustainability initiatives and programs. The trajectory of successes began with the imaginative vision of its mayor at the time, Jamie Lerner, whose goal was to uplift the city's inhabitants through projects involving increased green space and preservation, car-free streets, and public transit that today is used by 75% of the city's population. Reducing flood devastation was an initial goal; Curitiba accomplished this by surrounding the city with green space that acts as a flood barrier. Since 1970, the city planted 1.5 million new trees and created 28 new parks. The city recycles 70% of its trash and possesses a well-designed, ultra-modern rapid transit bus system. Poverty rates are low, the city is clean, and residents have a pride of place. The Free University for the Environment in Curitiba provides no-cost training to residents about sustainability, ensuring that community knowledge, capacity, and know-how will be passed down to present and future citizens.[8]

According to the Arcadis 2022 Top 100 Sustainable Cities Index, the 20 cities highest-ranking in sustainability are in North America or Europe, with the exception of Tokyo, which came in at number 3. Often ranked as first in pan-Asian countries, Tokyo already had an impressive record when it announced in the summer of 2022 that it would revitalize an abandoned parcel of land near Tokyo Bay previously used during the 2020 Olympics. Tokyo plans to finish the new 1,000-hectare (2,471-acre) project by 2050. The new development can be imagined as a city-within-a-city and is nicknamed Tokyo-eSG, referring to its ambition to be the planet's first ESG city and a model for global city development. Tokyo-eSG is aimed at achieving a circular economy based on renewable energy, with buildings and public transportation that have zero emissions.[9]

The Arcadis 2022 Index honored six African cities in the top 100: Cairo (Egypt), Cape Town and Johannesburg (both in South Africa), Nairobi (Kenya), Lagos (Nigeria), and Kinshasa (Democratic Republic of Congo). These six cities scored well on 32 criteria, including air quality, drinking water access, biodiversity protections, climate resilience, waste management, and energy efficiency.[10]

Cultural traditions are intertwined with ecosystems, as they have been for as long as humanity has been on Earth. On the island of O'ahu, Hawai'i, community members and three grassroots nonprofit organizations are working together to restore an 800-year-old traditional fishpond that provides food and spiritual sustenance as the center of the community itself over generations. As with all of the case studies featured here, this site is in a region undergoing ever-increasing population pressure—in this case, mainly tourism and an influx of outsiders. Over time, the watershed where the pond is located has suffered pollution from agricultural and development runoff, sewage, sedimentation and other sources. The volunteers are removing invasive plants and re-establishing the native taro root, a traditional food that is part of Hawaiian polyculture. In 2017, the United States National Oceanic and Atmospheric Administration (NOAA) designated the coastal area of He'ieia as National Estuarine Research Reserve (NERR). The designation added protection of 1,385 acres of unique biodiversity within ecologically critical upland ecosystems, wetlands, reefs, and seagrass beds and invaluable cultural and historic areas. All of the restoration activities are contributing to making He'ieia and her people more resilient, including climate change.[11]

Taking a deep look at the many scales, settings, and interpretations of sustainable communities around the world, several common principles, values, or goals can be identified as key attributes. Increasingly frequently, these attributes are not implemented alone but are integrated holistically into community projects. What follows is a short, preliminary list of examples, to which communities both can and must add.

9.3 Common Themes and Elements in Global Community Planning for Sustainability

1. Common themes and elements[12]
2. Climate resilience[13]

3. Nature-based solutions to increase community resilience[14]
4. Energy efficiency
5. Food accessibility and affordability; local food initiatives
6. Health care accessibility/affordability
7. Inclusivity[15]
8. Repurposing and renovation[16]
9. Resilience to storms
10. Walkability; public transportation including equity and accessibility for disabled citizens
11. Waste elimination (circular economy, zero waste)
12. Water conservation, water quantity, water quality

9.4 References

1. *IUCN Global Standard for Nature-based Solutions: A User-Friendly Framework for the Verification, Design, and Scaling up of NbS* (1st ed.). Gland, Switzerland, 2020. https://portals.iucn.org/library/sites/library/files/documents/2020-020-En.pdf (accessed November 1, 2022).
2. Eduardo Blanco, Kalina Raskin, and Philippe Clergeau. 2022. Reconnecting Neighborhoods with Ecosystem Functioning: Analysis of Solutions from Six International Case Studies. *Sustainable Cities and Society*, 77:103558.
3. (1) Exploration Architecture. www.exploration-architecture.com/projects, (accessed November 1, 2022); and (2) (Sustainable Cities Initiative) Ciudades Sostenibles: Biomimicry: Designing Cities According to Nature. https://blogs.iadb.org/ciudades-sostenibles/en/biomimicry/ (accessed November 1, 2022).
4. Eduardo Blanco, Kalina Raskin, and Philippe Clergeau. Reconnecting Neighborhoods with Ecosystem Functioning: Analysis of Solutions from Six International Case Studies. Op Cit.
5. Ibid.
6. C40 Reinventing Cities Competition. www.chicago.gov/city/en/depts/dcd/supp_info/c40-reinventing-cities-competition.html (accessed November 16, 2022).
7. Pascal Sabino. 2022. *Garfield Green Affordable Housing To "Counteract Pressures Of Displacement" in East Garfield Park*. Block Club Chicago. https://blockclubchicago.org/2022/02/22/city-commission-approves-garfield-green-affordable-housing-to-counteract-pressures-of-displacement-in-east-garfield-park/ (accessed November 16, 2022).
8. (1) Oliver Satrvi. 2021. *The Most Sustainable City in Latin America*. Curitiba, Brazil. www.greenzine.org/post/the-most-sustainable-city-in-latin-america-curitiba-brazil (accessed November 16, 2022), (2) B. Boselli. 2016. *The Latin American Green City Index: Assessing the Environmental Performance of Latin America's Major Cities. 2010.* https://biblioteca.cejamericas.org/bitstream/handle/2015/2686/

Latin_American_Green_City_Index.pdf?sequence=1&isAllowed=y (accessed November 16, 2022), and (3) United Nations Sustainable Development Goal 11: Make Cities and Human Settlements Inclusive, Safe, Resilient, and Sustainable, The 2030 Agenda for Latin America and the Caribbean. https://agenda2030lac.org/en/sdg/11-sustainable-cities-and-communities (accessed November 16, 2022).

9. (1) Tokyo-eSG. The acronym ESG stands for environmental, social, and governance—a shorthand buzzword for principled, stakeholder-driven investing that prioritizes positive environmental, social, and governance characteristics in addition to financial aspects in order to maximize sustainability and minimize risk but also to standardize the framework for reporting on sustainability in terms of investing. From the standpoint of leaders, citizens, and planners of innovative sustainable projects, strong ESG characteristics attract venture capital from a generation of investors looking to make a difference with their investment power. For more, see this example: Witold Henisz, Tim Koller, and Robin Nuttall. 2019. Five Ways that ESG Creates Value. *McKinsey Quarterly*, 4. www.mckinsey.com/capabilities/strategy-and-corporate-finance/our-insights/five-ways-that-esg-creates-value (accessed November 16, 2022) and (2). Zero-emission. Stefan Ellerbeck. 2022. Tokyo Wants to Build a Future-Proof City: Here's How. *World Economic Forum* (September 21, 2022). www.weforum.org/agenda/2022/09/japan-tokyo-city-sustainable-future/ (accessed November 16, 2022).
10. Waste management and energy efficiency. (1) Benoit-Ivan Wansi. 2022. Africa: At Least Six Cities in a Global Ranking of 100 Sustainable Cities. *Afrik21: News on the Green Economy, The Environment, and Sustainable Development in Africa* (June 16, 2022). www.afrik21.africa/en/africa-at-least-six-cities-in-a-global-ranking-of-100-sustainable-cities/ and (2). United Cities and Local Governments of Africa. www.uclga.org (accessed November 16, 2022).
11. Holly V. Campbell and A. M. Campbell. 2017. Community-Based Watershed Restoration in He'eia (He'eia ahupua'a), O'ahu, Hawaiian Islands. *Case Studies in the Environment*, 1(1):1–8. https://doi.org/10.1525/cse.2017.sc.450585.
12. Repurposing and Renovation. *C40Reinventing Cities Could Be a Reference Here, to Repurposing and Renovation, Same Reference as the Chicago Case Study*. www.c40reinventingcities.org (accessed November 16, 2022).
13. (1) UN Habitat for a Better Future: Cities and Climate Change Initiative. https://unhabitat.org/initiative/cities-and-climate-change-initiative (accessed November 6, 2022) (2) UN Habitat for a Better Future: World Cities Report. 2022. *Envisaging the Future of Cities*. https://unhabitat.org/wcr/ (accessed November 6, 2022), and (3) (Zibi, Ottawa, and Ontario, Canada. The Waterfront City. https://zibi.ca (accessed November 6, 2022).
14. An impact evaluation framework to support planning and evaluation of nature-based solutions projects. https://ora.ox.ac.uk/objects/uuid:3ecfc907-1971-473a-87f3-63d1204120f0/download_file?file_format=pdf&safe_filename=EKLIPSE_Report1-NBS_FINAL_Complete-02022017_LowRes_4Web.pdf&type_of_work=Report (accessed November 6, 2022).
15. Inclusivity: Office of Environment & Heritage, NSW Government, Australia. www.environment.nsw.gov.au/resources/communities/110563-Building-Sustainable-Communities.pdf (accessed November 7, 2022).
16. (1) C40 Reinventing Cities Competition. www.chicago.gov/city/en/depts/dcd/supp_info/c40-reinventing-cities-competition.html (accessed November 16, 2022) and (2) Stefan Ellerbeck. 2022. Droughts are Getting Worse Around

the World; Here's Why and What Needs to Be Done. *World Economic Forum* (August 12, 2022). www.weforum.org/agenda/2022/08/drought-water-climate-un/ (accessed November 9, 2022).

9.5 Discussion Questions

1. How can we incorporate a more integrated approach to sustainable planning—in other words, build in social, environmental, and economic goals from the beginning? Hint: refer back to the IUCN's eight standards for nature-based solutions and the UN Sustainable Development Goals.
2. How do you interpret the observation that the local *is* the global?
3. Should every country have its own model of Curitiba, Brazil's Free University for the Environment? What are some support roles that education (including higher education) can play in sustainable community planning?
4. Can you think of additional common themes and elements in global community sustainability planning to add to the list? Can you locate examples in your country or around the world?
5. How might cities and countries share their sustainability initiatives and projects in order to promote cross-community learning?

Conclusion

> You do not need to know precisely what is happening, or exactly where it is all going. What you need is to recognize the possibilities and challenges offered by the present moment, and to embrace them with courage, faith, and hope.
>
> —Thomas Merton[1]

Communities may sometimes feel that their efforts are just a tiny drop in the bucket, but when it comes to sustainability, the local is simultaneously the global. Moreover, the energy and foresight of collaborative local, state, and regional efforts can continue to advance and innovate even during times when the efforts of national governments are stalled by political conflict, policy inertia, or the paralysis of administrative decision-making. This profound phenomenon—the mirroring of the local and the global—has always been true as a touchstone of our collective freedom to shape our awareness, understanding, surroundings, futures, and legacy.

Having focused, contemplated, researched, written, and taught for decades about themes for sustainability across our vocational and avocational paths, the convergence of the book's themes in terms of climate resilience became ever clearer by the end of our book project. All of the goals of sustainability are, as many others have pointed out, "things we must do anyway," but as they are integrated into everyday life, these goals add up to a revolutionary shift in how we more cohesively see (and think anew about) ourselves, our dwellings and towns, our livelihoods, and the way we live in a community—both locally and globally.

At the time of writing this conclusion to our volume, the magnitude of both the damages done and the scale of global effort needed for repair are staggering. Examples of recent events include catastrophic monsoon flooding in Pakistan and India (causing the deaths of thousands of citizens, including children, and costing untold damages estimated to be in the tens of billions of dollars) and the ongoing drought in East Africa and Europe.

Veteran policy experts and thought leaders, including John Kerry, emphasize the need for rich nations to support vulnerable nations in the Global South not only with technological solutions but also with financial commitments, which he sees as a *proactive* investment in human beings—in our common humanity—and which he distinguishes from calls for reparations or mechanisms to address past harms (retroactive). Note, that the two paths of action are not mutually exclusive.[2]

Toward the very end of the conference, the delegates to team COP27 in Sharm el-Sheikh, Egypt in November 2022 finally did agree to a historic fund for nations inequitably bearing the brunt of climate change impacts.

The agreement is referred to as the "loss and damage fund." Egypt's foreign minister and COP27 president deemed the agreement "an answer to cries of anguish and despair" from developing nations.[3]

The steps we take to envision change and then implement projects on the ground (one after another) can generate cohesive, integrated (systems) co-benefits that include clean air and water, reduced energy consumption (and therefore reduced emissions), cooler dwellings, buildings, and cities. The list goes on: expansion of green spaces that promote relaxation and lower stress, accessibility of locally grown fresh food, places to gather and share conversation, music, play, nature, culture and the arts, happier and healthier citizens (including children and seniors), a more positive social and public health context that supports human dignity and equality, and greater meaning and fulfillment.

As we gather the skills and creativity to collaboratively move our communities toward greater sustainability, we are a force of nature; as you work in your region know that within the dynamic, worldwide holistic energy your own group is a vital part—you are not alone; you are part of the new powerful design coming into view.

References

1. Thomas Merton. 1966. *Conjectures of a Guilty Bystander*. Image Books, Doubleday, New York, NY., 328 pp.
2. Justin Worland. "Moral Obligation." John Kerry Says Developed Countries Need to Ramp Up Help for Growing Climate Losses. *Time* (October 28, 2022). https://time.com/6225834/john-kerry-loss-and-damage-climate-interview/ (accessed November 9, 2022).
3. Fiona Harvey, Nina Lakhani, Oliver Milman, and Adam Morton. 2022. COP27 Agrees Historic "Loss and Damage" Fund for Climate Impact in Developing Countries. *The Guardian* (November 20, 2022). www.theguardian.com/environment/2022/nov/20/cop27-agrees-to-historic-loss-and-damage-fund-to-compensate-developing-countries-for-climate-impacts (accessed December 13, 2022).

Glossary

A

adaptation: a process of change through which a species becomes sustainably suited to its habitat. Over the past two decades with regard to human response to climate change, this term is frequently prioritized over the term "mitigation." However, presently communities and governments aim to undertake actions that both *mitigate* (lessen, as in achieving net-zero carbon emissions) and help people *adapt* to changes already occurring and forecast to worsen (increased heat, severe storms, floods, fires).

adaptive management: a structured process that allows acting within uncertain conditions based on the best available science, closely monitoring and evaluating outcomes, and re-evaluating and incrementally adjusting decisions in ways that maximize opportunities to learn from experience.

albedo effect: the electromagnetic radiation reflected back into space by the white surface of a growing ice sheet; *albedo* is Late Latin for "whiteness," from the Latin *albus*, "white."

B

benefits: something that produces good or helpful results or effects, thereby promoting well-being.

berm: an artificial ridge or embankment.

biomimicry: a design principle that strives to create architecture, urban design, and planning elements based on nature. Biomimicry seeks sustainable projects that solve problems and feature net positive benefits for people and the environment. See *Rethinking the Future: Biomimicry in Urban Design*: www.re-thinkingthefuture.com/architectural-styles/a2531-biomimicry-in-urban-design-a-topic-of-your-choice.

built environment: human-made structures, features, and facilities viewed collectively as an environment in which people live and work.

buffer: a natural, restored, or designed and planned ecological feature such as a riparian forest or vegetated strip between an ecosystem such as a river, stream, or wetland to separate the ecosystem from surrounding uses, such as agriculture heavily developed or industrial areas. The buffer replaces and maintains necessary ecosystem services, such as native plants for pollinators and wildlife, filtration of pollutants and sediment from precipitation runoff, and aesthetic beauty.

C

carbon cycle: the process in which carbon atoms continually travel from the atmosphere to the Earth into and through organisms and then back into the atmosphere.

carbon sequestration: the process of capturing and storing atmospheric carbon dioxide.

charrette: a participatory planning process that involves a wide diversity of collaborating stakeholders and interdisciplinary specialists and/or contributors. The benefits of a charrette include the following: (1) it can give quick results, (2) it can boost creativity, and (3) participants can observe their project from an integrated viewpoint because it correlates all relevant disciplines to create a plan that balances environmental issues with land use, economic considerations, and transportation. (The World Bank. https://urban-regeneration.worldbank.org/node/40 [accessed September 28, 2022])

circular economy: a conscious behavior intended to keeps materials, products, and services in circulation for as long as possible, out of the waste stream (US Environmental Protection Agency; www.epa.gov/recyclingstrategy/what-circular-economy); a model of production and consumption that involves sharing, leasing, reusing, repairing, refurbishing, and recycling existing materials; a systems solution framework that tackles global challenges like climate change, biodiversity loss, waste, and pollution (Ellen MacArthur Foundation, https://ellenmacarthurfoundation.org/topics/circular-economy-introduction/overview).

climate: a measure of how the atmosphere "behaves" over considerably large geographical areas and relatively long periods of time. On the other hand, weather is more local, seasonal, or short-term.

climatic cycle: the cyclic changes in weather patterns in a geographical area over time.

co-benefits: the positive effects an action (such as an on-the-ground sustainability project), policy, or measure aimed at one ecological objective might have as a multiplier toward other ecological objectives, including social welfare. The goal is to efficiently and simultaneously maximize co-benefits.

community: in the human sense, a group of people living in and using a common area with various degrees of cohesion, cooperation, and integration.

cost-benefit analysis (CBA): the exercise of comparing the costs and benefits of taking a particular course of action. While historically CBA was narrowly conducted using money as its metric, modern CBA includes the flexibility of applying other values as metrics, although among the values used actual currency may be an important one (according to the Harvard Business School, see https://online.hbs.edu/blog/post/cost-benefit-analysis).

cycle: a series of events that are regularly repeated in the same order.

cycling: to occur in or pass through a cycle; to move as if in a circle.

D

data: information empirically derived, known or assumed as facts that can be used in calculating, reasoning, or planning.

depaving: the act, usually with volunteers, of identifying unnecessary, unused, and excessive pavement (concrete, for example, in schoolyards or unused parking lots) at locations within a community, then removing the pavement and replacing it with rain gardens, microforests, and other green areas. The co-benefits of depaving include absorption of precipitation (instead of runoff or flooding), cooling a neighborhood, providing shade if trees are planted, providing clean air, and storing carbon (see impermeable/permeable).

diversity: the relative degree of abundance of species of plants and animals, functions, communities, habitats, or habitat features per unit of area.

drainage: the system or process whereby water or other liquids are drained from a particular location. A watershed is also sometimes referred to as a drainage.

dynamic: characterized by or tending to produce continuous change. Compare and contrast with turbulence, stochastic systems, and other terms describing the behavior of natural systems (terrestrial, aquatic, atmospheric, and marine).

E

ecological: an adjective that identifies a relationship between living organisms and their non-living physical environment.

economic valuation: provides a tool to assist with comparing various options involving ecosystem services as an attempt to make the value of services visible and accounted for along with monetary values. On the other hand, *non-economic* valuation refers to environmental goods and services, such as clean air and water, and healthy populations of fish and wildlife that are not traded in markets. Their economic valuehow much people would be willing to pay for them—is not necessarily revealed in market prices unless a citizen valuation study was undertaken.

ecosystem: the standard reference is Odum (1969), which classifies ecosystems as having biological structure, physical structure, energy flows, and material flows (Eugene P. Odum, "The Strategy of Ecosystem Development," 164:262 *Science*, April 18, 1969).

ecosystem services: processes of natural systems that directly or indirectly benefit humans by enhancing social welfare.

ecosystem values: the kind of beneficial functions performed by such ecosystems as wetlands to the abundance, diversity, and habitats of fish and other wildlife.

equality/equity: the criterion of making sustainability accessible and affordable to people of all nations, incomes, neighborhoods, and backgrounds without regard to their location, culture, race, or financial status. Equity is closely related to climate justice; the concept and value of equity are critically important to address the unequal environmental and climate impacts burden borne by poorer neighborhoods, historically and through today, globally.

evapotranspiration: the contribution of evaporated water to the atmosphere from all sources: dew, rain or snowfall, or water bodies on Earth's surface, from the capillary fringe of the groundwater table, from the uptake of water by plants which then respire vapor to the atmosphere.

F

flood attenuation: an ecosystem service provided by floodplains, wetlands, engineered stormwater retention ponds, and other features in the landscape that naturally or are designed to help slow and hold water to achieve flood prevention and management and to conserve water.

flood: an overflowing of a vast amount of water beyond its normal limitations, especially over normally dry land.

floodplain: a plain bordering a stream or river that is subject to flooding.

flood retention: the temporary or permanent withholding of floodwaters in a designated basin during rain or flood events, when the risk of flooding is highest. Flood retention is a characteristic of floodplains, wetlands, rain gardens, and bioswales.

G

geographic information system (GIS): a computer system for capturing, storing, checking, and displaying data (mapping) related to positions on Earth's surface at a variety of scales.

grant: a sum of money given by a government or other organization for a particular purpose.

grantee: the person to whom a grant or conveyance is made.

grantor: the party in a transaction who conveys ownership of an asset.

grant report: a regular, required summary report reflecting the goals and outcomes of the original grant request. It details how the received funds were used by the organization to which they were given and the impact they created on the community.

greenhouse effect: the way heat is trapped close to Earth's surface by greenhouse gases (GHGs). These heat-trapping gases can be thought of as a glass shield wrapped around Earth, keeping the planet warmer than it would otherwise be.

greenhouse gases (GHGs): gases such as carbon dioxide and methane that have the property of absorbing infrared radiation. Radiation is energy (heat) that comes from a source, such as the sun, and travels through space at the speed of light. GHGs include water vapor, ozone, carbon dioxide, methane, nitrous oxide, and chlorofluorocarbons. Emitted from Earth's surface and reradiating back to its surface, these gases contribute to the greenhouse effect. Sources of GHG include transportation, power plants, livestock, manure, sewage, fertilizer and other aspects of agriculture, manufacturing, and many other sources.

greenspace: land that is partly or completely covered with grass, trees, shrubs, or other vegetation, including greenways, parks, golf courses, community gardens, and cemeteries.

H

habitat: the sum total of environmental conditions of a specific place (food, shelter, water, space, and privacy) occupied by a plant, an animal, or a population of such species.

habitat corridor: is a natural connection that provides food and shelter for a variety of wildlife and helps with juvenile dispersal and seasonal migrations by allowing or maximizing a more-or-less contiguous space for movement among fragmented habitats and their isolated populations, thereby promoting increased genetic diversity. In the United States, contemporary habitat corridors can increasingly include human-engineered structures that help wildlife migrate or otherwise move across the landscape, avoiding mortality on freeways and highways, to relocate to adjacent habitat areas.

Happiness Index: the World Happiness Report (WHR) annually ranks the world's nations based on responses from their citizens about how happy they are. You can review your country's ranking in the WHR at (1) https://worldhappiness.report and (2) The United States Happiness Index, which similarly ranks each state but according to research metrics rather than survey results of citizens' own evaluations (the US report for 2022: https://wallethub.com/edu/happiest-states/6959).

hardscaping: any of the non-living, harder elements in your landscape design, such as concrete, rocks, bricks, pavers, stone, and wood.

I

impermeable: material through which liquids and gases cannot pass, a prime example being pavement, whether concrete or asphalt and the like.

L

landscape: a physiographic unit that is capable of sustaining several populations of a species.

M

master-planned communities: impressive examples of resilience to disturbances, damage, or destruction from catastrophic storms, flooding, and fire. The details of these communities' planning criteria and strategies, and stories of survival are compelling, but master-planned communities are not affordable to the great majority of people. These communities possess practical, contemporary knowledge-based planning lessons that must be extended to the greatest extent possible to all communities. See "Ciara Nugent, Climate-Proof Towns Are Popping Up Across the US But Not Everyone Can Afford to Live There," October 28, 2022, *TIME*, https://time.com/6225970/climate-proof-towns-extreme-weather/.

mitigation: making up for the loss of something, also used frequently as synonymous with "improvement." For example, the term "wetland mitigation" refers to the action of creating, restoring, or enhancing wetland habitat often in response (in the US) to a federally permitted action that caused the loss of wetlands. Climate mitigation refers to reducing or eliminating greenhouse gas (GHG) emissions.

migration: a periodic movement away from or back to a given area.

monitoring: refers to regular, periodic objective observation, measurement, and recording of key factors (for example, water quality, temperature, fisheries and wildlife, or other important criteria or benefits) for evaluating part of a plan in progress or one that has been completed.

N

nature-based solutions: actions to protect, sustainably manage, and restore natural and modified ecosystems that address societal challenges effectively and adaptively, simultaneously benefiting people and nature. Nature-based solutions address societal challenges through the protection, sustainable management and restoration of both natural and modified ecosystems, benefiting both biodiversity and human well-being. Nature-based solutions are underpinned by benefits that flow from healthy ecosystems. They target major challenges like climate change, disaster risk reduction, food and water security, biodiversity loss, and human health and are critical to sustainable economic development. Definition of the IUCN, www.iucn.org/our-work/nature-based-solutions Examples of nature-based solutions include bioswales and rain gardens. See also: IUCN Global Standard for Nature-Based Solutions: A user-friendly framework for the verification, design, and scaling up of NbS, First Edition, Gland, Switzerland: IUCN, (2020), which contains further international case studies, https://portals.iucn.org/library/sites/library/files/documents/2020-020-En.pdf.

nitrogen (N_2): a nonmetallic element constituting nearly four-fifths of the air by volume; a colorless, odorless, almost inert gas. It occurs in various minerals and in all proteins. Compare with reactive nitrogen species resulting from human and natural processes.

nutrient cycling: the circulation of elements, such as nitrogen and carbon, via specific pathways from the nonliving to the living portions of the environment and back again.

O

ordinance: a local law enacted by a county or municipal authority.

oxygen: a colorless, odorless, tasteless gaseous element constituting 21% (by volume) of the atmosphere. It combines with most elements, is essential for plant and animal respiration, and is required for nearly all combustion and combustive processes.

P

permeable: having pores or openings that permit liquids and gases to pass through.

pollutants: harmful materials, or contaminants, released into the environment.

pollution: the introduction of harmful materials, termed pollutants, into the environment.

population: a group of individuals that are interfertile and that regularly contribute germ cells to the formation of fertile offspring.

private property: ownership by private parties, essentially anyone or anything other than the government; may consist of possessions like real estate, buildings, objects, or such intellectual property as copyright.

public property: property owned by the government (or its agency) rather than by a private individual or company. In other words, it belongs to and is used by the public at large, including many parks, streets, sidewalks, libraries, schools, and playgrounds.

R

rain gardens and bioswales: green areas for receiving and filtering rain. These areas reduce contamination, runoff, and flooding and conserve water, and they are examples of nature-based solutions.

regulation: a rule or directive made and maintained by a government or institutional authority.

resilience: the flexible ability of systems and plant and animal species to successfully adapt to changes in the environment (or for humans, adaptation also to life challenges).

restoration: to return something to a "perceived" former position or condition; restoration of biodiversity, restoration of ecosystem functions and services.

right-of-way: a legal right to access private land, such as a road.

riparian zone: an area identified by the presence of vegetation that requires free or unbound water or conditions, moister than normally found in the area. Riparian zones are adjacent to creeks, streams, rivers, and other freshwater systems and are frequently enhanced or restored in order to provide shade, lower water temperature and increase water quality and habitat for native plants, fish, and other wildlife, including invertebrates.

rule: a duly enacted regulation that implements the goals of a statute or a law. Compare this with *guidance*, a written document issued by a public agency or other authority to provide the regulated public with helpful information toward solving a problem.

runoff: the draining away of water (or substances carried in it) from the land's surface that ultimately reaches a stream (see *stormwater*). Compare this with the velocity, volume, and contamination of runoff over paved (impervious) environments with runoff across covered lands (fields, slopes, crops, forests).

S

soil: earth material so modified by physical, chemical, and biological agents that it will support rooted plants. One may identify soil as any one of six types: sandy, silty, peaty, chalky, loamy, or clay, all with different properties.

slope: a surface on which one end or side is at a higher level than another.

SMART goals: specific, measurable, achievable, relevant, and time-bound goals connected to your project's purpose.

species: a unit of classification of plants and animals consisting of the largest and most inclusive array of sexually reproducing and cross-fertilizing individuals that share a common gene pool.

statute: a law (sometimes called an *act*) enacted by the legislative branch of a government.

stormwater: surface water in an abnormal quantity as a result of the runoff of heavy rain and meltwater from hail or snow. If captured in bioswales or stormwater wetlands, stormwater can be naturally filtered and conserved.

sustainable: the ability to be maintained something at a certain rate or level.

sustainability: fulfilling the necessities of current generations without compromising the requirements of future generations. Sustainability can only be accomplished by ensuring compliance with nature's biophysical processes, which are the long-term foundation of economic growth and social well-being.

U

urban agriculture: the production, processing, and marketing of food and other agricultural products within the geographical limits of urban and suburban settings within a metropolitan area, making food accessible to area citizens.

urban heat island: an urban or metropolitan area composed of concrete, steel etc. buildings, roads, and other highly concentrated, human-created structures (with limited, intermixed vegetation). This concentration of structures absorbs the sun's heat and becomes an "island" that, at night, then gradually re-emits higher temperatures than such natural landscapes as grasslands, forests, and water bodies.

United States Environmental Protection Agency (USEPA) Building Blocks Program: a collaborative initiative with the United States Department of Housing and Urban Development and the United States Department of Transportation. The goals include environmental protection, promotion of equitable development, and provision of tools to address climate change (www.epa.gov/smartgrowth/building-blocks-sustainable-communities). Note: in the United States, the EPA and several other federal and state agencies have grant funding for communities to apply for initiatives and collaborations that address and improve environmental and watershed conditions.

W

water cycle: beginning in the ocean, water evaporates into the air and is transported by the wind as a vapor to the land, where it falls as rain, snow, or ice (the latter two returning to a liquid as the temperature warms). There the liquid either goes into the soil or flows downhill into streams and rivers, both below- and aboveground, on its way back to the ocean, where the cycle began. Evapotranspiration is the process by which water is transferred from the land to the atmosphere by evaporation from the soil and other surfaces and by transpiration from plants through pores in their leaves.

weather: what the atmospheric conditions are like in localized areas over a relatively short period of time. By contrast, *climate* is a measure of how the atmosphere "behaves" over considerably large geographical areas and relatively long periods of time.

wetland: an area, such as a marsh or swamp, where a gradation of wet conditions prevails in the soil all year varying periods of time, including the growing season. "Treatment wetlands" are constructed as systems that use natural processes involving wetland vegetation, soils, and their associated microbial assemblages to improve the quality of water.

Appendix: Selected Further Resources for the Reader

Adewole, Taiwo and Associates, Waste Management Towards Sustainable Development in Nigeria: A Case Study of Lagos State. 2009. *International NGO Journal*, 4: (4):173–179. https://academicjournals.org/article/article1380901268_Adewole.pdf.

ASHRAE, founded in 1894, describes itself as "a global society advancing human well-being through sustainable technology for the built environment." The site contains case studies. www.ashrae.org/communities/chapters/ashrae-chapters/ashrae-community-sustainability-projects.

Benfield, Kaid, What Does a "Sustainable Community" Actually Look Like? *The Atlantic*, March 14, 2011, www.theatlantic.com/national/archive/2011/03/what-does-a-sustainable-community-actually-look-like/72376/, and Natural Resources Defense Council's All-In Cities Project, focused on reducing climate impacts and addressing income inequality, www.nrdc.org/issues/build-sustainable-cities.

1:24 Beyond Housing, Grassroots Social Sustainability in St. Louis, Missouri, United States, www.beyondhousing.org. This is a model program for community transformation based on multi-pronged initiatives driven by citizens and supported by a framework known as "Ask-Align-Act."

Business for Social Responsibility (BSR), Sustainable Communities Case Studies, provides tools to help companies develop community-beneficial philanthropic investments; international case topics covered include circular economy, climate change, collaboration, conflict minerals, consumer products, energy and extractives, environment, ethics and governance, sustainability management, sustainable communities, women's empowerment, and others. www.bsr.org/en/topics/case-studies/Sustainable-Communities.

Climate Chance, Sustainable Community Project (SCP), Ghana, www.climate-chance.org/en/best-pratices/sustainable-community-project-scp/, by Green Africa Youth Organization, seeks to implement "a circular economy model of material use and replicate this model across the continent . . . The sustainable community Project envisions a community where: production is green, waste is utilized to generate income and households are well educated on proper waste management. The project is creating a community-driven proper waste management structure through an incentive-based approach. The project was piloted at the Adansi South District of Ghana through Public-Private Partnership, Public Education, and Stakeholder Capacity Building."

Climate Change Resources, American Planning Association, includes resources on climate change, energy, green infrastructure, and hazard mitigation, www.planning.org/resources/climatechange/#Climate%20Chang.

Depave is a nonprofit organization in Portland, Oregon, USA, whose mission is to empower communities through urban regreening. While limited to Portland, its concepts, philosophy, project types and community involvement could be replicated around the world.

Ellerbeck, Stefan, These Four Cities are Encouraging People to Protect Biodiversity: Here's How, World Economic Forum, Centre for Nature and Climate (November 29, 2022), www.weforum.org/agenda/2022/11/biodiversity-cities-wildlife-nature/. This article is about a global collaboration between the World Economic Forum (WEF) and the government of Colombia for the development of an urban model designed in harmony with nature with the participation of citizens, cities, and businesses. The collaboration is called BiodiverCities by 2030 (see www.weforum.org/communities/biodivercities-by-2030).

Groundwork USA links communities in the pursuit of equity and sustainability. They work with communities throughout the US. https://groundworkusa.org.

Inter-American Development Bank, an organization with the mission to improve the lives of communities throughout Latin America and the Caribbean, www.iadb.org/en, and its Ciudades Sostenibles (Sustainable Cities) Program, https://blogs.iadb.org/ciudades-sostenibles/en/.

Institute for Sustainable Communities, https://sustain.org, is an international nonprofit organization. ISC has "worked on over 100 projects in 30 countries throughout the world," currently focusing on the largest GHG emitters and those most impacted by climate disruption. This site includes descriptions of ISC projects but not complete case studies.

Joslin, Sierra, 9 Inspiring Sustainable Community Examples, *gb&d Magazine* (Green Building and Design), June 15, 2021, https://gbdmagazine.com/sustainable-community-examples/.

Komeily, Ali, and Srinivasan, Ravi S., A Need for Balanced Approach to Neighborhood Sustainability Assessments: A Critical Review and Analysis, *Sustainable Cities and Society*, 18 (2015):32–43.

Locke, Jayna, Six Traits of a Sustainable City (With Examples), Blog, *Digi International*, June 17, 2021, www.digi.com/blog/post/sustainable-city; also includes a list of the top ten most sustainable cities in the world and discussion of the benefits of sustainable cities.

Macdonald, Fiona, review of a book about reusing abandoned buildings (with photos and examples). August 17, 2022, *BBC Culture Series*, "Designed to Last,": www.bbc.com/culture/article/20220809-designed-to-last-ten-of-the-worlds-most-ingenious-buildings (title of the book: *Building for Change: The Architecture of Creative Reuse*, by Ruth Lang, Gestalten [2022]).

McLendon, Russell, People in Portland Planted Trees. Decades Later, a Stunning Pattern Emerged, *Science Alert*, November 25, 2022. www.sciencealert.com/people-in-portland-planted-trees-decades-later-a-stunning-pattern-emerged. This article explains how a tree-planting initiative in Portland, Oregon, ended up generating multiple economic and public health benefits years later.

RainWise Program, Seattle, Washington, United States. Seattle's population is approximately 734,000, and its annual rainfall is 37 inches. RainWise is a publicly funded rebate program that could be replicated in other cities. The program, hosted by King County Wastewater Treatment Division and Seattle Public Utilities, in partnership with five sister organizations (and growing), promotes green stormwater infrastructure (GSI) by helping homeowners establish cisterns, rain gardens, or both to capture rainwater that would otherwise end up in the streets and a municipal stormwater collection system. The goals include flood prevention, landscape aesthetics, and stored water for dry-season irrigation and prevention of combined sewer overflow (CSO) events during extreme precipitation that lead to pollution of area waterways. Cisterns and rain gardens are tools to intercept and filter runoff. The ultimate goal is to capture 700 million gallons annually; the initiative currently intercepts 465 million gallons per year. The website features case studies including Venema Creek neighborhood, which intercepts 70% of stormwater from an 80-acre area of pavement and keeps the contaminated stormwater from entering Venema Creek. See https://700milliongallons.org.

Serve-Learn-Sustain, Case Studies on Sustainable Communities, Georgia Tech University, https://serve-learn-sustain.gatech.edu/tool-category/case-studies.

Sharifi, Ayyoob, and Akito Murayama, A Critical Review of Seven Selected Neighborhood Sustainability Assessment Tools, *Environmental Impact Assessment Review*, 38 (2013):73–87.

Sustainable Development Goals Fund of the United Nations (SDGF), Case Studies—Achieving SDGs (www.sdgfund.org/case-studies), is an invaluable online database featuring best practices for achieving a sustainable world, implementing the United Nations' comprehensive 17 Sustainable Development Goals. All 17 goals are necessarily interrelated (reflecting the three pillars of ecological, social, and economic sustainability discussed in this book); thus, the goals are meant to be integrated to synergistically work together holistically. You can learn about the 2030 Agenda for Sustainable Development and the 17 goals here: https://sdgs.un.org/goals and https://sdgs.un.org/goals#goals. *Goal 11 specifically pertains to sustainable cities and communities.* The case studies are from all over the world. While this site notes that the Sustainable Development Goals Fund is operationally closed, an announcement dated February 15, 2022, indicates that the Joint Sustainable Development Goals Fund expanded its portfolio and selected five countries' proposal submissions to fund (from over 100 nations). The 2022 countries selected are Kenya, Madagascar, North Macedonia, Suriname, and Zimbabwe. https://unsdg.un.org/latest/announcements/joint-sdg-fund-doubles-its-portfolio-114-million-catalytic-impact-investments.

Sustainable States Network (United States) lists members of a network of state programs that help thousands of communities improve their quality of life via concrete action steps on sustainability, climate, and clean energy. www.sustainablestates.net/state-programs.

The World Bank, Sustainable Cities and Communities, www.worldbank.org/en/topic/sustainable-communities and the related Sustainable Cities Blog, https://blogs.worldbank.org/sustainablecities.

Index